Beyond Sovereign Territory

BORDERLINES

Beyond Sovereign Territory
The Space of Ecopolitics

THOM KUEHLS

BORDERLINES, VOLUME 4

University of Minnesota Press

Minneapolis

London

Published by the University of Minnesota Press
111 Third Avenue South, Suite 290
Minneapolis, MN 55401-2520

Book design by Will Powers
Set in Sabon and Officina types
by Stanton Publication Services, Inc., Saint Paul, Minnesota
♲ Printed on recycled paper (50% recycled/10% postconsumer)

Library of Congress Cataloging-in-Publication Data

Kuehls, Thom.
    Beyond sovereign territory : the space of ecopolitics / Thom Kuehls.
        p.   cm. — (Borderlines ; v. 4)
    Includes bibliographical references and index.
    ISBN 0–8166–2467–4 (hardcover). — ISBN 0–8166–2468–2 (pbk.)
    1. Environmental policy.   2. Sovereignty.   I. Title.   II. Series:
Borderlines (Minneapolis, Minn.) ; v. 4.
GE170.K84   1996
363.7—dc20   95–31240

Contents

Acknowledgments

There are a number of people I would like to thank for their contributions to this work. Students at the Johns Hopkins University, Goucher College, and Weber State University have worked through a number of the issues in this work with me in courses on Environmental Politics. Weber State University provided me with a grant to travel to Vancouver Island to research the ecopolitical struggles over the logging of the ancient temperate rain forests there. The Clayoquot Sound Biosphere Project and the Friends of Clayoquot Sound both were generous in allowing me access to their copious files. Jane Bennett, David Campbell, William Connolly, William Corlett, Kara Shaw, Michael Shapiro, Cindy Weber, and an anonymous reader for the University of Minnesota Press all read the entire manuscript at one stage or another and provided excellent criticisms and comments. Kara Shaw in particular deserves a special note of thanks. Her assistance throughout this process was invaluable. As an undergraduate at the College of Wooster, it only took a couple of classes from Mark Weaver my first term to convince me that political theory was the direction I wanted to go with my studies. Mark has remained a valued friend and adviser. Finally, I would like to acknowledge Amelia Jenkins, Dave Kuehls, Ernest Kuehls, Iris Kuehls, and John Kuehls; there is a bit of each of them in this book.

Introduction: Brazil

"The Amazon is ours," declared José Sarney, president of Brazil, in 1989 in a statement titled "Our Nature." "After all, it is situated in our territory."[1] Sarney's statement, directed in part at the "great powers or international organizations . . . that would come to dictate to us [Brazil] how to defend what is ours to defend,"[2] is an expression of sovereignty. As such it fits into a powerful understanding of the space of politics: states do, or must, control the politics across their territories. The tropical rain forest of the Amazon River valley is said, by Sarney, to belong to Brazil because, geographically, it lies within the boundaries that mark the territory of the state of Brazil. Consequently, the sovereign authorities of the state of Brazil are *the* decision makers when it comes to the fate of this forest. But Sarney's statement is a tremendous oversimplification of the issue. And the terms of his statement, from "territory" to "our," are far from unproblematic.

In general terms, this sovereign territorial description of political space fails to contain politics along two basic lines: the inability of the space of sovereignty to contain the flows of political, economic, and ecological activity, and the extent to which both the territory and the population of sovereign states are constructed through practices that exceed the apparatus of state sovereignty. What I propose to do in the following pages is examine political space through a focus on the politics of ecology, or ecopolitics. What I will suggest is that descriptions of political space that isolate politics to particular institutions that contain sovereign authorities and are themselves contained within particular territories barely begin to capture the diversity of the space of (eco)politics. My examination of political space will follow two basic avenues: one involving a discourse of sovereignty, the other involving a discourse of "government," or "governmentality."

SOVEREIGNTY

Brazil's sovereignty over the Amazon rain forest has been challenged by politicians and environmentalists on the ecological grounds that the importance of this rain forest extends far beyond the territory of Brazil. The function of large tropical rain forests like the Amazon in producing oxygen, absorbing carbon, and generally regulating weather patterns around the globe begins to suggest the manner in which the Amazon does not lie (solely) within the territory of Brazil. Moreover, the clearing of this rain forest is contributing to the global increase of carbon dioxide, a principal "greenhouse" gas, in the earth's atmosphere. In 1988, for example, the burning of the Amazon rain forest produced almost one-quarter of the carbon dioxide released into the earth's atmosphere.[3] The CO_2 that is released into the atmosphere when the Amazon rain forest is cleared is not contained by the geographic boundaries of Brazil. Its effects may well be global. The ecology of the rain forest flows across the border lines of the sovereign territorial space of Brazil, perhaps extending the "territory" of the rain forest in ways that confound discourses of sovereignty.

As long as Sarney (and others) can present territory as a fixed, static entity, his claim of sovereignty over it can avoid certain ecologically problematic issues. However, when the ecology of areas is taken seriously, that is, when the manner in which various ecosystems slice across geopolitical boundaries is highlighted, territory becomes much less of a fixed, static entity and claims of sovereignty over it become increasingly problematic. When this happens, the politics within the boundaries of a sovereign territory can no longer be unproblematically rooted to that specific place.

"The physical effects of our decisions spill across national frontiers," the United Nation's World Commission on Environment and Development (WCED) declared in its 1987 report on the earth's ecology and economy.[4] If sovereignty is rule over a specific territory, but the physical effects of political decisions made in one sovereign territory spill over into other sovereign territories, is not the sovereignty of these various territories subverted? Sarney sought to reaffirm the sovereignty of Brazil by attempting to reduce the influence of "great powers or international organizations" on Brazilian politics. But the discourse utilized by these great powers and international organizations has very often also been the discourse of sovereignty.

Unable to think politics outside this discourse and space, when faced with problems that slice across the present boundaries of sovereign space, state actors have sought solutions through the extension of the space of sovereignty. "This authority must have power," exclaimed Norwegian prime minister Gro Harlem Brundtland at a global atmosphere conference at The Hague in 1989.[5] Calling for an appropriate United Nations entity to be given the power to "police the global atmosphere,"[6] Brundtland and other participants at that conference were calling for an extension of the space of sovereignty beyond the nation-state. They were calling for the creation of an institution with "the appropriate measures to enforce its directives."[7] Rather than question the inefficiency of the discourse of sovereignty to address the problem of global atmospheric pollution, Brundtland and others simply sought to extend its territory.

Neither Sarney's invocation of sovereignty nor Brundtland's is sufficient in the face of transversal ecological effects. The problem, I am arguing, lies not with the size of sovereign territories, but with the concept of sovereignty itself. Ecopolitics forces an engagement with a host of questions that challenge the otherwise unproblematic presentation of the space of sovereignty.

The territory of the sovereign state is not reducible to geopolitical boundaries. How the boundaries of territory and sovereignty are made must be examined. In other words: How is it that a particular area of the earth's surface becomes sovereign territory? Is it as simple as drawing lines around an area, controlling the people who live within these lines, and defending them and the area against potential intruders? Is that all that is necessary to create the "our" of Sarney's proclamation? What of the "our" of the WCED's statement? Do they connote the same thing? How does this "our" come into being?[8] Does it encompass "public" as well as private activities? Does sovereignty only rule over the former and not the latter, or does it pretend to rule over both? Is territory simply the geographic region upon which sovereignty is exercised? Or are various practices that exceed the realm of sovereignty necessary to create territories out of contingent and ambiguous lands and peoples?

GOVERNMENTALITY

When John Locke wrote in 1698 that "land that is left wholly to nature, that hath no improvement of pasturage, tillage, or planting, is called, as indeed it is: waste,"[9] he was formulating not just an economic argument about how land is made into property (and hence, territory), but an ethical argument about how persons ought to interact with the land. Through his notion of "waste land," Locke defended a "natural right" to property that excluded numerous orientations to land from this right. According to Locke, the "several Nations of the *Americans*" could not be said to own the land upon which they lived because they did not put the land to use.[10] Indeed, from Locke's perspective the vast continents of North and South America were made up of wastelands due to the non-European inhabitants not having put them to use. But Locke did not recognize the particularity of his notion of use. Hence the "natural right" to property he advocated was actually only present to people who labored upon the land in the particular manners deemed appropriate by Locke. Through this lens, the land in the Americas prior to the arrival of Europeans could be considered open to become the property of anyone who would come in and put his/her labor to it.

The implications of this conception of property for thought about sovereign territory might well be obvious. If the individual inhabitants of the Americas prior to the arrival of Europeans could not be said to own tracts of land because they did not "use" it, how could the several Nations of the Americas make any claim to sovereignty over the land? Sovereign territory, when read through Locke's defense of property rights, becomes an economically and ecologically specific concept. It must exhibit explicit signs of being worked in a particular manner in order to become sovereign territory.

Similarly, the population of a sovereign territory must be people who have adopted this specific orientation to the land. If the inhabitants of the Americas could not be said to be owners of property, nor sovereigns over a land, then Locke is also making a statement about the type of person capable of operating within the discourse of sovereignty. Rather than sovereignty being a simple exercise of a people over a territory, complex practices must take place to create the territories and peoples for the exercise of sovereignty. Thus, Brazil can only be a sovereign territory if it is populated by a particular people and if the land is utilized in a specific manner.

"Nothing will prevent us," proclaimed Brazilian president Getulia Vargas in 1940, "from accomplishing . . . the conquest and the domination of the great valleys of equatorial torrents, *transforming* their *blind force* and *extraordinary fertility* into *disciplined energy*."[11] Vargas's desire to transform the force and fertility of the Amazon rain forest into energy helped set the stage for the destruction we are currently witnessing there. From Vargas's vantage point, this transformation was necessary for the process of state building. To this extent, his statement can be read as an exhortation of sovereignty. For many political theorists, this is perhaps a strange claim.

The transformation of territory is not generally considered to be a problem of sovereignty. Sovereignty, Western political theory has traditionally taught, is about imposing a law across a territory and upon a people—the condition of the territory and the character of the people are "mere variables" in the practice of sovereignty.[12] But the political theorist operating within this framework of political thought is either unwilling or unable to see the problem of "government."

According to Jean-Jacques Rousseau, "government" is about providing for needs, while sovereignty is about authority.[13] The problem of "government," according to Michel Foucault, encompasses concerns that had been (and still are) "mere variables" to (perhaps) the majority of modern political theorists: "resources, means of subsistence, the territory with its specific qualities, climate, irrigation, fertility, etc."; and people with their specific qualities, "customs, habits, ways of acting and thinking, etc."[14] These issues are not independent of the exercise of sovereignty over a territory but are inextricably linked to it. Far from being given entities, sovereign territories (including their populations) are created entities. And the creation of sovereign territory carries with it profound ecological consequences.

As with the physical effects of decisions that slice across territorial boundaries, the "governmental" practices that operate to create sovereign territories are not contained by these sovereign spaces. Brazil cannot transform the Amazon rain forest into disciplined energy without "outside" help. The capital that drives the ongoing transformations of land into territory is global(izing). The conquest and domination of the Amazon rain forest by the state of Brazil requires the assistance of multinational corporations such as Goodyear, Nestle, Volkswagen, Borden; it also requires the assistance of international organizations such as the World Bank and the United Nations Food and Agriculture Organization.

The border lines that divide up the contemporary geopolitical realm into separate sovereign territorial states are pathetically porous. Thought concerning ecopolitics must take seriously this porosity. Ecopolitics takes place on border lines—not just border lines between states, but border lines between humans and nature, populations and territories, and sovereignty and government.

INTO THE AMAZON

Until quite recently, vast tracts of the Amazon rain forest were considered "empty" by the Brazilian government. Despite the fact that tens of thousands of people lived in these forests, these areas were deemed empty due to the absence of any signs of land occupation. The principal sign of occupation acknowledged by the state of Brazil was the clearing of the jungle.

In keeping with the impassioned plea of President Vargas in 1940 to put the Amazon rain forest to use, the Brazilian government initiated the Plan for National Integration in 1960, with the intent of "occupying" the Amazon rain forest as fast as possible. To facilitate this occupation, the state provided incentives for miners, ranchers, and other settlers to enter this area. Official estimates are that the population of the state of Amazonia in Brazil increased from one hundred thousand in 1960 to two million in 1970.[15]

Faced with this "Lockean" understanding of use and property, it should come as no surprise that the principal method of "preserving" particular areas has been to fence them in and keep people out. But this strategy has come under considerable criticism over the past decade for a number of reasons: one being that it assumes a certain static character to nature, that areas left alone will continue in their present state in perpetuity; a second being that it discounts other methods of use, methods that are far less destructive of natural systems than those that have principally been employed in the name of development.

The first critique is perhaps the most prevalent in the dominant discourses of global environmentalism. "Conservation," argues the former vice president for science at the World Wildlife Foundation, Thomas Lovejoy, "has traditionally been a static, defensive affair. The idea has been to put a fence around it and it'll stay that way . . . We've been managing the planet for a long time now, mostly by neglect. Now we simply have to learn how to intervene properly and to manage it positively."[16] For the people involved in the Minimum Critical Size of Ecosystems project, cosponsored by Brazil's National Institute for Amazonian Research and the U.S. branch of the World Wildlife Fund, the task is just that—managing positively. Aware that many parks and wilderness areas are deteriorating largely due to insufficient size, Lovejoy and others in this project feel that by participating in the deforestation of the rain forest they can help save it by identifying the appropriate size tract of rain forest that will be self-sustaining.

Despite the lush vegetation that marks tropical rain forests such as those found in the Amazon River valley, the soil is typically very poor. These rain forests do not live off of the soil as much as they live off themselves. Life comes to these rain forests through death: as older trees die, new plant life takes root in the fallen logs. When isolated into small enough patches, the rain forest can literally starve to death.[17] Thus, simply fencing in a patch of rain forest is not sufficient to preserve it. If the area is not large enough, the plants—and animals that depend on the rain forest for habitat—will inevitably die off.

"Understanding the implications of the fragmentation of tropical rain forests is essential to rational government planning of human settlements and land use . . . if development is to proceed with minimal disturbance of the natural ecosystem," wrote Rob Bierregard, et al., directors of this study.[18] If the equatorial torrents of this rain forest are to be turned into disciplined energy without eradicating the entire rain forest, knowledge of the characteristics of the rain forest is necessary. What the Minimum Critical Size of Ecosystems project presents is a "governmental" ecology. It offers a "green" way to transform the ambiguous space of the rain forest into the disciplined space of sovereign territory. Even the uncut areas will no longer be wild, for they will be intensely studied in order to further the knowledge of rain forest ecosystems in order that development can become even more environmentally friendly. Positive management of natural areas, rather than benign neglect, seems to be the watchword of this project.

An alternative method for saving the rain forest that involves leaving the forest standing and making use of it as a forest has recently received a great deal of attention. "Long-term exploitation of forest products," as opposed to rapid deforestation, ranching, or mining, has been argued to be both more profitable and more eco-logical. What "long-term exploitation" involves is labor-intensive, environmentally friendly harvesting of the nuts, fruits, and other economically valuable products of the forest by the inhabitants of the forest. But, as Charles Peters of the Institute of Economic Botany at the New York Botanical Garden (a proponent of this sort of use of the rain forest) maintains, the process is not dependent upon "Indians."

Speaking of one "long-term exploiter," Peters remarked:

> This guy is not an Indian. There's no lip disk. No body paint. This guy is wearing a Michael Jackson T-shirt and Adidas running shorts, but he still knows 100 species of fruit tree. And he's got that gleam in his eye. He not only knows how to find the fruit, he knows how to bring the fruit to market and he knows how to haggle.[19]

As with the territory, the people must also undergo a transforma-tion. If "long-term exploitation" is to work as a viable economic and ecological activity, the operators presumably cannot be "Indians," those people that Locke claimed demonstrated no signs of use upon the land. Although "Indians" had taken the fruit from the trees of the rain forest for centuries, it is important that this harvester not be an "Indian"; it is important that this harvester have "that gleam in his eye" that sees the rain forest as an economic opportunity.

As with the Minimum Critical Size of Ecosystems project, "long-term exploitation" is an example of "governmental" ecology. It aims at transforming both the land and the people in order to meet the needs of the society (the global society, in these cases, since the eco-nomic activities discussed here are not contained by the economics of Brazil), albeit in a proposed ecological manner. But the principal element at work still follows the rationality of Locke's notion of property. In order for the rain forest to be put to use, the individuals doing the harvesting and the harvesting process itself must adhere to certain standards. The nuts and fruits that are harvested must contribute to the overall economy of the state and perhaps even the globe. In that manner, the individuals engaged in the harvesting are contributing to these economies as well. Neither can escape the transformative aspect of governmentality.

The issues of sovereignty and governmentality, glimpsed through this brief look at the ecopolitics of the Brazilian rain forest, will frame the discussions of chapters 2 and 3, respectively. Chapter 4 is an examination of the field of ecological political thought in light not just of the discussions of the space of politics in chapters 2 and 3 but also of the discussion of eco-ethics that takes place in chapter 1. In chapter 5 I return to "Brazil," this time "Brazil of the north," in an attempt to flesh out further the theoretical discussions of the previous chapters.

The work before you is primarily an exploration of the space of ecopolitics. But exploring this space, recognizing its complexities, gaining an appreciation for its contingencies should not be an end in itself. It is not my contention that once politics moves to this space and abandons the space of the sovereign state, the ecological problems we face will be solved. Politics already takes place here. Allow me to repeat that: politics already takes place in this nonsovereign space. What is needed, aside from ways of thinking about politics in this nonsovereign space is an ethic to accompany this thought.

In chapter 1 I attempt to articulate my own eco-ethic, drawing on the work of Friedrich Nietzsche and recent examples of what has come to be called "chaos theory." This eco-ethic is not unique to the space of ecopolitics I am exploring. It is only one of many possible eco-ethics that could inhabit this space. But it is intended to engage what I consider to be the dominant ethical orientation at work in this space today: a "Lockean" ethical orientation that privileges a certain utilitarian relationship between humans and nonhuman nature, where the earth is largely interpreted as given to humans.

1

Natures, Ethics, and Ecologies

Nature left undisturbed, so fashions her territory as to give it almost permanence of form, outline and proportion.   G. P. MARSH 1864

God is dead; but given the way of men, there may still be thousands of years in which his shadow will be shown.   F. NIETZSCHE 1882

Change now appears to be intrinsic and natural at many scales of time and space in the biosphere.   D. BOTKIN 1990

"The true ecologist," writes Anna Bramwell in her history of ecology, "could not have arisen until the middle of the last century."[1] The reason for this, Bramwell and other "historians of science" conclude, is that "earlier epochs all . . . see the earth as man's unique domain precisely because of God's existence."[2] Ecology, understood here as a perspective, an ontology, or a worldview wherein humanity's actions pose potential problems for the earth, could not have existed as long as the earth was felt to have been not only created by God for humanity, but cared for by God for humanity as well. As long as God was in charge of seeing to the condition of the earth, its condition could never become a problem. In order for the ecologist to be born, God had to die.

The premodern world was filled with a god that was intimately linked to the world's condition. It was during that time, Bramwell contends, that

1

both religious *and* natural theology were impregnated with the idea of a God-centered world. When science took over the role of religion in the middle of the 19th century, the belief that God made the world with a purpose in which man was paramount declined. But if there was no purpose, how was man to live on the earth? The hedonistic answer, to enjoy it as long as possible, was not acceptable. If man had become God, then he had become the shepherd of the earth, the guardian, responsible for the *oeckonomie* of the earth.[3]

On this reading, the birth of the ecologist is inextricably linked with the death of god. As god's role in maintaining the condition of the earth declined, a space was opened up for the ecologist to emerge. For many historians of the field of ecology, this emergence began just over a century ago.

THE PROBLEM OF MAN AND NATURE

Just over a hundred years ago, George Perkins Marsh wrote *Man and Nature*, a work widely considered to be one of the first environmentally conscious, or ecological, works.[4] Although the term ecology would not be coined by German biologist Ernst Haeckel until shortly after the publication of *Man and Nature*, Marsh's work has been argued to represent a "new perspective" in Western thought.[5] Marsh was one of the first Western thinkers to suggest that 'man'[6] could disrupt or destroy nature not simply on a local scale, but on grand scales, both temporally and spatially.

Living in the last decade of the twentieth century, with nuclear power, toxic waste, global deforestation, air that is unhealthy to breathe, global warming, holes in the ozone layer, and so on, such a thought might appear obvious. In the West in the mid-nineteenth century, however, the thought that humans might be able to radically alter nature was new. The newness of this thought can be seen in the response Marsh's editor had to the title Marsh originally suggested for his 1864 work. "Is it true?" Marsh's editor replied when Marsh suggested "Man the Disturber of Natural Harmonies" as the title for his work.[7]

Although Marsh's definition of 'man' as the disturber of natural harmonies was, arguably, new to the nineteenth century, much of what accompanied it in his text was not. "Nature left undisturbed," he argued, "so fashions her territory as to give it almost unchanging permanence of form, outline, and proportion."[8] This belief in the unchanging permanence of nature, with respect to form, outline, and proportion, dominated Western thought into the nineteenth century.[9]

Moreover, in keeping with the thinking of his era, Marsh's nature did not evidence this permanence of form, outline, and proportion by chance. Nature was created, created specifically with 'man' in mind. "Through the night of aeons," Marsh wrote, "[nature] had been proportion[ed] and balanc[ed], to prepare the earth for [man's] habitation, when, in the fullness of time, his Creator should call him forth to enter into its possession."[10] Neither Marsh's description of the permanence of nature's form nor his claim about the place of man in nature broke significantly from mainstream nineteenth-century Western thought on the subjects.

What *was* revolutionary about *Man and Nature* was its suggestion that 'man' can, and does, disrupt the harmonies of nature, harmonies that have been fashioned by God. "Wherever [man] plants his foot, the harmonies of nature are turned to discords," Marsh maintained.[11] While many in the nineteenth century trumpeted humanity's newfound power over nature, Marsh sounded a note of caution. Despite the recognition of human beings' ability to affect natural systems, the prevaling opinion of the time was that 'man' still could not be destructive to the earth on a large scale. A faith accompanied nineteenth-century technological advancement, a faith that 'man' was only playing out the role God intended for him to play. Marsh, however, argued that 'man' was no longer playing his intended role, and thus had become a destructive force on the planet.

Nature had been proportioned and balanced specifically for 'man''s existence, according to Marsh. Beginning with this assumption, Marsh pursued what was/is perhaps the obvious question: where did things go wrong?[12] The problem, he argued, was that "man ha[d] too long forgotten that the earth was given to him for usufruct alone, not for consumption, still less for profligate waste."[13] While Marsh held that usufruct did allow for "a certain measure of transformation of terrestrial surface, of suppression of natural, and stimulation of artificially modified productivity," he felt that "this measure man ha[d] unfortunately exceeded."[14]

In this regard, Marsh is not all that far from Locke. As I will argue, Locke did propose what could be considered an ecological provision to his theory of appropriation. Where Marsh differs from a Lockean ethical orientation is not on the general or theological relationship between 'man' and earth, but on the specific role 'man' was to follow, the level of activity he was divinely ordained to engage in.

'Man,' a "free moral agent, working independently of nature,"[15] Marsh maintained, having forgotten what he was created for, was now turning the harmonies of nature to discords. 'Man''s free will, the same characteristic that gave him his power over nature and made him "of more exalted parentage, and belong[ing] to a higher order" than the rest of "existences,"[16] also gave him the capacity to destroy nature's harmonies. Yet it also gave him the capacity to reorient his behavior to the natural harmonies of the world. This was the calling of Marsh's eco-ethic, to make 'men' realize that human activity had gone astray and hence to get them to make the appropriate alterations to bring their activities back in tune with nature.

What made Marsh's thesis revolutionary was not his conception of nature as regular, harmonious, and unchanging, nor his view that nature was created (created, moreover, with 'man' in mind), nor even that 'man' has the free will to deviate from God's intended plan. What made Marsh's thesis revolutionary was his belief that "nature did *not* heal herself" after 'man' had transformed or disfigured "her."[17] The nature that God had made for 'man' could be permanently altered by 'man.' God's creation was vulnerable, and, moreover, God was not there, presumably, to continually retune nature's harmonies.

Reading *Man and Nature* in light of Bramwell's presentation of the ecological perspective, the emergence of a new role for 'man' upon the earth is clearly evident. Marsh calls upon 'man' to become the ecological shepherd of the earth, to take responsibility for its *oeckonomie*. But Bramwell also speaks of a loss of purpose to the world, and a loss of purpose in which 'man' was paramount. This element of Bramwell's formulation of the ecological problematic is not as easy to find in Marsh's text.

Bramwell claims that the ecological problematic arose from a decline in the belief that God made the world with a purpose (a purpose in which 'man' was paramount). 'Man,' faced with a purposeless world, was forced to forge a new role for himself. "If there was no purpose, how was man to live on the earth?"[18] But the world in which Marsh constructed his ecological ethic was far from purposeless. This was a world that had been fashioned with an almost unchanging permanence of form, outline, and proportion. This was a world that, throughout the eons, had been proportioned and balanced for 'man''s occupation of it. Marsh (as well as many contemporary ecologists) required such a world to give his ecological ethic its force.

If God is dead, if there is *no* purpose to the world, how is it that wherever 'man' plants his foot the harmonies of nature are turned to discords? If it were not for the harmonious permanence of Marsh's world, what standards could exist by which to judge 'man''s actions? If truly there is no purpose to the world, it would seem that the preconditions for the ecological problematic as articulated by Marsh would not be available. Where would ecology be without a purposeful world? A demand for ontological reassurance seems to accompany this field of thought. If God is dead, from whence comes this ontological reassurance, this world still replete with purpose?

A European contemporary of Marsh's interrogated this demand for ontological reassurance, not just with respect to humanity's relationship to nature, but across much of the field of Western philosophical, political, and social thought. "God is dead," Nietzsche has become famous for writing. "But," he continued, "given the way of men, there may still be thousands of years in which his shadow will be shown."[19]

THE SHADOW OF GOD

Nietzsche's thoughts on the shadow of God provide an excellent entry into the purpose(s) of Marsh's ecology. For Marsh, and for those of us who might act on his recommendations, God may as well be dead. As Lowenthal pointed out in his 1965 introduction to *Man and Nature*, Marsh's lesson was that nature does not heal "herself" after 'man' transforms or disfigures "her." God is not there to repair the damage that we might bring to the earth. God is no longer responsible for the (future) condition of the earth. Yet, what might be referred to as God's shadow still resides in Marsh's ecological thoughts.

While Nietzsche is perhaps most (in)famous for uttering the words "God is dead," I will argue that Nietzsche was far more concerned with the lingering presence of God's shadow. "We distort Nietzsche when we make him into the thinker who wrote about the death of God," contemporary French philosopher Gilles Deleuze maintains. "For Nietzsche, this is an old story."[20]

New Struggles.—After Buddha was dead, his shadow was still shown for centuries in a cave—a tremendous, gruesome shadow. God is dead; but given the way of men, there may still be thousands of years in which his shadow will be shown.—And we—we still have to vanquish his shadow, too.[21]

To hear that God is dead may shock. But, as Deleuze contends, Nietzsche presents this as old news, something that has already transpired and is no longer worthy of lengthy consideration. There are new struggles to engage. "—And we—we still have to vanquish God's shadow, too."

Marsh's ecological perspective exists within the shadow of God. Although God is dead, nature still exhibits the characteristics of a divine creation and 'man' still exists in a divinely created relationship with nature. Nature still stands forth with its almost unchanging permanence of form, outline, and proportion. 'Man' still occupies an ontologically privileged position. Nietzsche, engaged in new struggles, attacks both this representation of nature and this representation of 'man''s position with respect to it. He attempts to draw nature out of God's shadow, to present it in a different light, and in doing so, he challenges the human conceit that maintains that nature was somehow created with us in mind.

RETHINKING MAN AND NATURE

In the aphorism following "New Struggles" Nietzsche issues a warning: "Let us beware."[22] On the one hand, Nietzsche warns us of believing that the world is a machine. "It is certainly not constructed for one purpose, and calling it a 'machine' does it far too much honor."[23] In this passage Nietzsche points to one ground of twentieth-century ecology: a mechanistic nature, a nature that normally runs with machinelike efficiency, yet can be tampered with by humans and also repaired by human mechanics (ecologists).[24]

Once the machinations of nature are discovered, the effects of human actions can be accurately measured. What actions are destructive and what actions are beneficial becomes a matter of proper calculation. As a machine, the world is opened up to what Nietzsche calls a "scientific" interpretation, an interpretation that reveals the mechanistic regularity, order, proportion of the world.

Yet, "a 'scientific' interpretation of the world...," Nietzsche asserts, might therefore still be one of the *most stupid* of all possible interpretations of the world, meaning that it would be one of the poorest in meaning. This thought is intended for the ears and consciences of our mechanists who nowadays like to pass as philosophers and insist that mechanics is the doctrine of the first and last laws on which all existence would be based as on a ground floor. But an essentially mechanistic world would be an essentially *meaningless* world. Assuming that one estimated the *value* of a piece of music according to how much it could be counted, calculated, and expressed in formulas: how absurd would such a "scientific" estimation of music be! What would one have comprehended, understood, grasped of it? Nothing, really nothing of what is "music" in it![25]

The mechanization of the world, an attempt to create an ordered and regular world following the death of God, has perhaps, Nietzsche suggests, created an essentially meaningless world. What can it comprehend of "nature"? But the mechanistic interpretation does serve a purpose. It restabilizes the world in the absence of God. The divine watchmaker may be dead, but God's watchlike world continues ticking with mathematical precision. If we can learn how it works, then we can repair what we've done wrong, as well as orient our activities to smoothly coincide with the gears and wheels of this nature.

While a mechanistic worldview can provide powerful arguments for environmentalists, from my Nietzschean ecological perspective, it still must be vanquished. In order to, what?—to finally give "meaning" to the world?

Nietzsche attacks the mechanistic representation of the world as one manifestation of the shadow of God. But does he offer his own representation of the world in response? Does he affirm one world as he attacks another? Where does Nietzsche turn in the face of his onslaught against mechanistic philosophy? Does he re-turn perhaps?

Nietzsche insists that he does indeed seek a "return" to nature, "although it is not really a going back but an *ascent*—up to the high, free, even terrible nature and naturalness where great tasks are something one plays with, may play with."[26] Contemporary environmental ethicist Max Hallman reads in this "return" an endorsement of an organic conception of nature. On Hallman's reading, Nietzsche's nature "must be seen as a living, growing, decaying process, and sometimes, perhaps even as a living being."[27] Having rejected the reading of Nietzsche as the thinker who "affirms the technological domination of the world,"[28] Hallman offers perhaps the only other reading available to him. If Nietzsche rejects mechanistic mastery, he must endorse organic attunement. Indeed, Hallman goes on to argue that Nietzsche's environmentalism prepares the way for humanity to be "finally able to live fully in the natural world."[29] Although Hallman's reading of Nietzsche provides a much-needed critique of the antiecological mastery reading of Nietzsche, ultimately I find his argument that Nietzsche embraced some type of organicism unconvincing; for Nietzsche offers another warning:

> Let us beware of thinking that the world is a living being. Where should it expand? On what should it feed? How could it grow and multiply? We have some notion of the nature of the organic; and we should not reinterpret the exceedingly derivative, late, rare, accidental, that we perceive on the crust of the earth and make of it something essential, universal, and eternal, which is what those people do who call the universe an organism . . . This nauseates me.[30]

Nietzsche warns of another possible manifestation of God's shadow, perhaps the leading conception of nature among environmentalists: nature as organism.[31] The drive to universalize, to eternalize what Nietzsche calls "some notion of the nature of the organic," only serves, in the end, the same *purpose* as the mechanistic world. It gives nature the appearance of harmony, consistency, unchangeability—a shadowy appearance. It takes one element of an ambiguous nature and solidifies it into *the* nature. It faces the prospect of a world beyond the death of God by remaining in the shadows. It continues to present the world as something predisposed to us. After all, if the world was not designed, somehow, with us in mind, why should we believe that we can attune to its harmonies, or "live fully" within it?

Nietzsche's attack on the twin representations of organism and machine constitutes a principal element in a reconceptualization of nature. His thoughts do not end with this destruction of old ideals, old metaphors, old values, old ways of life. He is interested in creation as well, and "for the game of creation . . . a sacred 'Yes' is needed."[32]

"The total character of the world, however, is in all eternity chaos—in the sense not of a lack of necessity but of a lack of order, arrangement, form, beauty, wisdom, and whatever other names there are for our aesthetic anthropomorphisms."[33] Nature is neither organism nor machine. These idols have been vanquished (at least in the thoughts of Nietzsche). Nature is different; nature is, in all eternity, chaos. To put it another way: "Nature is no model! It exaggerates, it distorts, it leaves gaps. Nature is *chance*."[34] Contemporary French philosopher Jean Granier, in an essay on Nietzsche's conception of chaos, argues that in Nietzsche's conceptualization, nature "has no grounds, nor reasons."[35] It cannot be the machine or organism that many environmentalists wish it to be. It is far too ambiguous to be contained by either conceptualization.

Nietzsche's claim is not that there is no consistency or stability to nature. His claim is that the *total* character of the world is *in all eternity* chaos, not in the sense of a lack of necessity, for, according to Nietzsche, "there are only necessities."[36] It is these necessities that hold nature together, if you will. The presence of certain elements in nature provide for the existence of other elements; certain causes produce certain effects. But far too often, we see in these necessities the eternal presence of order, arrangement, form, beauty, wisdom . . .

"What? Do we really want to permit existence to be degraded for us like this . . .?"[37] Such insistences "divest existence of its *rich ambiguity*," Nietzsche complains.[38] Such degradations take the form of an ecology grounded in mechanistic principles, for instance, where the ambiguity is squeezed out of nature, where nature is reduced to mathematical formulations and Euclidean shapes. This reduction eradicates difference through an insistence that nature be one way and shall not be another.

Nietzsche's insistence that the world is, in all eternity, chaos, is an insistence that our categories, our interpretations, do not exhaust the world. Nature is always different. Nietzsche is acutely aware that interpretation is necessary. That is not the focus of his attack. His charge is against those who would insist that their interpretation is the true one, those who would deny the rich ambiguity of existence in order that the world they describe be the true world. For most of the history of Western philosophy, Nietzsche maintains, "the 'true world' has been constructed out of contradiction to the actual world."[39] Western thought has consistently posed the "true world" as a world of permanence, a world of unity, a world of harmony, a world of ideals, heaven, God. In doing so, the actual world, according to Nietzsche, has been denied: the world of flows, diversity, "discordant concordances,"[40] bodies, earth . . .

As a means of recognizing this actual world, Nietzsche pursues the question "What is nature, for me?"[41] The particularity of this question carries with it tremendous ecological importance. Challenging the universality of many attempts to define nature, Nietzsche asserts that a thing would be defined only after *all* beings had answered the question "What is that?"[42] Nietzsche's attack on universalist interpretation is not limited to differences within the species of human being, although I would not want my following argument to suggest that these differences are not crucial to developing an eco-ethic. What I am interested in setting out here is Nietzsche's conception of nature and his conception of humanity's relationship to it, and for this reading it is necessary to focus on the Nietzschean possibility "whether . . . all existence is not essentially engaged in *interpretation*."[43] The potential for differences in interpretation when this possibility is entertained becomes virtually limitless. Nietzsche's argument is not that interpretation is not possible or needful. By hammering home the point that *perspective* is implicit in interpretation, he problematizes it, attacking the human conceit that would hold our interpretations as *the* necessary ones, or *the* true ones.

By reconceptualizing nature in this manner, Nietzsche necessarily problematizes the standing of humanity on earth. Not only do our interpretations fail to grasp the earth in its entirety—even though many of us continually insist that they do—but they are only one among billions of possible interpretations, coming from all types of life forms. Nietzsche's perspectivism breaks radically with the bulk of Western thought in extending the power of interpretation to nonhuman life forms. But, since Nietzsche asserts that "there is [no] necessity for men to exist,"[44] one ought to take seriously the possibility that our interpretations are not the only ones and hence that the world was not designed with 'men' in mind. This is not to say that the world is completely hostile to 'men.' If that were the case, certainly our species would have perished long ago. There are certain necessities to the world that have enabled human beings to survive, but that is not the same as suggesting that our existence is necessary.

Marshian environmentalists are unable to entertain the possibility that the world was not designed with 'man' in mind. 'Man' must be able to locate his place within the world in order for their environmentalisms to operate—this is the case for both mechanistic and organic breeds.[45] Without this place, they think, no limits can be placed upon human actions at all. The high, free, frightful nature, the nature that is, in all eternity, chaos, which Nietzsche presents, could not house an eco-ethic from their perspective. But these environmentalisms fear to tread out in the light, out beyond the safety of God's shadow.

The ecological problematic as it was articulated by Marsh responded to the death of God. It recognized a problem with how 'man' was relating to nature. It struggled to write a new ethics for 'man''s relationship to nature. However, it did so from within the shadow of God. Nature remained a created object, created, moreover (explicitly for Marsh, but implicitly for many who would follow him), for 'mankind.' And 'man' remained an object divinely ordained to put nature to his use. In short, it responded to a Lockean orientation to nature without significantly challenging the assumptions upon which that orientation was based.

A WORLD OF CHAOS?

The time may be upon us, however, when even God's shadow is being vanquished. Thus, the time may be upon us when a different eco-ethic needs to be articulated. One that is not dependent upon a nature that exhibits the regularity of a machine, or provides the comfort of an organic home that can be returned to. One that radically alters how we can think ourselves with respect to nature. The 'man' and nature that Nietzsche presented contains strong affinities with 'men' and natures that are emerging across the sciences and humanities—natures of "chaos."[46]

These natures do not demonstrate a permanence of form, unity, proportion; they need not be created with humanity in mind; they need not be predisposed to our being nature's shepherd, guardian, or master, thus creating new humanities. These new natures create new problems. Just as Marsh and others of his era had to struggle to construct new values and new ethics within a world confronted by the death of God, we must struggle with a world in which God's shadow may be disappearing. We face a new problematic.

"We are so accustomed to the laws of classical dynamics that are taught to us early in school," argue physicist Ilya Prigogine and chemist Isabella Stengers, "that we often fail to sense the boldness of the assumptions on which they are based."[47] The "laws of nature" that a great many of us have been taught are based upon assumptions of unity, permanence, and simplicity. These assumptions no longer command the way they once did. "Our vision of nature," Prigogine and Stengers suggest, "is undergoing a radical change toward the multiple, the temporal, and the complex."[48]

Classical science insisted upon a universal regularity to nature. Nature was thought to be smooth, ordered, consistent, straight-lined. A glorious union of nature and Euclidean geometry was ordained. "The shapes of classical geometry are lines and planes, circles and spheres, triangles and cones," writes James Gleick in his popular work *Chaos*. "They represent a powerful abstraction of reality, and they inspired a powerful philosophy of Platonic harmony."[49]

It is worth pausing here a moment to think seriously the implications of this claim. The geometric conceptions of space that dominate our thinking are powerful abstractions of reality. When Supreme Court Justice Sandra Day O'Connor remarked in a 1993 redistricting case that the new district in question was unconstitutional because of its "bizarre" shape, she was making an interpretation about political space influenced by Euclidean geometric abstractions.[50]

> Fractal geometry challenges this abstraction. It presents space as "rough, not rounded, scabrous, not smooth. It is geometry of the pitted, pocked, and broken up, the twisted, tangled, and inter-twined."[51] As with Euclidean geometry, fractal geometry is also an abstraction. But its presence helps to erode the belief in the realism of a world that is straight-lined and smooth.

> Fractal geometry is not simply jagged-lined Euclidean geometry. The challenges it poses to classical conceptualizations of space are not just with respect to the lines that may surround certain areas or objects. Not only are the shapes and spaces of fractal geometry irregular (whatever that means), they are also twisted, tangled, and intertwined. Fractal geometry forces us to think of spaces as interacting, perhaps even as dynamic.[52]

> The space of the territory of the United States, then, becomes difficult to solidify from a fractal perspective. Can it be isolated to a static area bounded on all sides by geopolitical lines? Or do flows of people, capital, air, water, and so on impact on the space of the United States, continually altering it? Can U.S. space be thought of in isolation from Canadian space, Mexican space, Japanese space?

Our geometric models are not the only ones that have been dominated by visions of regularity. Mathematical models have long operated under the same constraints. "No good ecologist ever forgot that his equations were vastly oversimplified versions of real phenomena," Gleick claims. "The whole point of oversimplifying was to model regularity. Why go to all that trouble just to see chaos?"[53] Regularity was the purpose. But the procedure was self-fulfilling. If regularity was the desired outcome, and the models were finely tuned to present a picture of regularity, the world would be regular.

Predator-prey models were built upon this love of the regular. As the deer population rises, so will the wolf population, we were taught: half a phase later. There is a regularity to the relationship that is not necessarily there, but can be found in the models. The incredibly complex set of variables (rainfall, plant life, insect populations, viruses, etc.) that would have to go into equations set up to predict deer and wolf populations might lead the ecologist to see a chaotic relationship rather than a regular one.[54] But Western thought has long insisted that it models reality, and that reality evidences these characteristics of regularity.

"The wealth of reality," however, Prigogine and Stengers maintain, "overflows any single language, any logical structure."[55] There should be no reason why natural relationships ought to correspond to our insistences. Pursuing the implications of such a thought seriously, instead of merely mentioning it in a passing phrase, could be earth-shattering, if you will. What if we thought the powerful implications of a world that is complex, contingent, chaotic?

I am not insisting that the world is as certain aspects of chaos theory suggest it is. I am attempting to highlight a perspectivism that emerges from Nietzsche's thought, a radical notion of interpretation that forces us to challenge numerous ways of thinking that have all too long been hidden within the shadows. My object is not finally to shine the light of truth upon the world. That would be very un-Nietzschean, as I read him. "'Is it true God is present everywhere?' a little school girl asked her mother; 'I think that's indecent'—a hint for philosophers! One should have more respect for the bashfulness with which nature has hidden riddles and iridescent uncertainties."[56]

What I am pursuing through this Nietzschean reading of aspects of chaos theory are the various ways in which thought about nature is being challenged. These "chaotic" thoughts are not limited to conceptions of space and mathemtical formulas, but have to do with time as well. Marsh had no real appreciation for the contingency of time. His text speaks of a nature that will continue to function perfectly, if left to itself, as if it were a great clock created by a god. But "nature as an evolving, interactive multiplicity," Prigogine and Stengers suggest, ". . . resist[s] its reduction to a timeless and universal scheme."[57] Time (re)emerges through chaos theory and brings with it a new set of problems.

Time, here, is not a linearly progressive element, but a radical concept that disrupts notions of order and regularity by bringing with it "the great theme of contingency."[58] For paleontologist Stephen Jay Gould, the concept of contingency is not to be placed within that restrictive dichotomy of rigid determinism and relativistic randomness. Contingency stands forth as a different possibility, "off the line," if you will, "a thing unto itself, not the titration of" the other two.[59] A contingent world is, recalling Nietzsche, a world of necessity, not determinism; a world of necessity, not randomness. The presence of human life on this planet is contingent; it is due to certain necessities.

"Wind back the tape of life to the early days of the Burgess Shale; let it play again from an identical starting point, and the chance becomes vanishingly small that anything like human intelligence would grace the replay."[60] The radicalness of this thought should not be brushed aside: "the chance becomes vanishingly small that anything like human intelligence would grace the replay." The presence of human beings on this planet is neither a deterministic nor a random event. Chance is involved. But each chance event leads to certain necessities, necessities that in turn enabled human life to evolve.

As we delve into the ideas of this theory of chaos, we are confronted with a new understanding not only of nature, but of ourselves. Put in terms of an accompanying model of evolution, Gould argues: "If humanity arose just yesterday as a small twig on one branch of a flourishing tree, then life may not, in any genuine sense, exist for us or because of us."[61] Our privileged place in the grand scheme of things begins to crumble. In nature beyond the shadow of God, there appears to be no necessity to human existence.

If, as Gould suggests, the history of life can no longer be thought of as a "conventional tale of steadily increasing excellence, complexity, diversity,"[62] which places humans at the apex of evolution, but is rather to be thought of as a "copiously branching bush, continually pruned by the grim reaper of extinction ["who, perhaps, is Lady

Luck in disguise" (Gould 1989, p. 48)], not a ladder of predictable progress,"[63] it becomes increasingly difficult to sustain theories that place humanity in an ontologically privileged position. The contingency of time problematizes our standing as—if not the pinnacle of creation—the pinnacle of evolution. "Perhaps we are only an afterthought," Gould suggests, "a kind of cosmic accident."[64]

For many environmental ethicists, and ethicists in general, this thought is a particularly troubling one. If there is no necessity to human existence, if we do not have a purpose on this planet, there appear to be no limitations to human action. From this position, vanquishing the shadow could not possibly offer a positive ecological stance. But this is only because these eco-ethicists cannot envision an ethic beyond the shadow of God. They feel that if no natural (read theological) limitations are placed on human beings—that is, if we do not have a specific role to play—we will be unleashed upon the earth to destroy it. As though we are not already engaged in this project from within the shadow of God.

As I will argue later in this chapter, I believe a Nietzschean ecological ethic can exist that understands nature to be chaos and sees no necessity to human existence. This will not be an ethic of egoistic hedonism, or of anything goes, but instead will be an ethic that highlights the difference and ambiguity of nature and sees a "role" for humans to not divest existence of these characteristics.

CHAOS AND ECOLOGY

The science of ecology has, over the last few decades, engaged the concepts of contingency, complexity, and chaos—concepts that, moreover, problematize the founding principles of ecology. "Until the past few years," suggests Daniel Botkin, "the predominant theories in ecology either presumed or had as a necessary consequence a very strict concept of a highly structured, ordered, and regulated, steady-state ecological system . . . Change now appears to be intrinsic and natural at many scales of time and space in the biosphere."[65] The belief in a nature that, if left alone, will maintain permanence of form, outline, and proportion is being seriously challenged by the results of ecological research. The view that dominated the science of ecology "during its first hundred years . . . along with its application in the management of living resources, . . . nature [as] a snapshot . . .

fixed in time and space, as constant as the environment of a modern scientific laboratory,"[66] is being undermined.

In *Discordant Harmonies*, Botkin points to several examples of failed policies in environmental management that were based upon beliefs in the timeless constancy of nature, beliefs that nature left untouched by human hands would remain in a static state of balance forever. The notion of a "static landscape, [existing] like a single musical chord sounded forever," Botkin insists, "must be abandoned."[67] Change occurs with or without human interference.

In many ways, Botkin's ecology is a direct response to that of George Perkins Marsh. It was Marsh who claimed in 1864 that "nature left undisturbed, so fashions her territory as to give it almost permanence of form, outline and proportion."[68] Botkin's claim that change is natural, at many scales of time and space, flies in the face of Marsh and a century's worth of ecologists and wildlife managers who followed him. Botkin is aware of the philosophical or ethical difficulties inherent in the shift from a constant nature to a continually changing nature:

> As long as we could believe that nature undisturbed was constant, we were provided with a simple standard against which to judge our actions, a reflection from a windless pond in which our place was both apparent and fixed, providing us with a sense of community and permanence that was comforting.[69]

Botkin speaks directly to the Nietzschean problem of God's shadow. Ecologies of constancy operate in this shadow, comforting us in the presence of a world where God is dead. But changing our reflection does not mean the end of ecology. Botkin argues that rethinking both natural processes and our relationship to them is essential for the practice of ecosystem management.

If change is intrinsic and natural (at many scales of time and space), the principles of the science of ecology must be reevaluated. Where change used to be the problem for ecologists, now it describes the ecological problematic. If nature is change, the ecologist's task must also change. No longer can s/he hold to beliefs in the unquestioned good of static landscapes;s no longer can s/he unproblematically attack any changes that might take place on the earth—whatever the scale of time or space. Life itself is dependent upon change.

Life, moreover, can no longer be viewed as a characteristic of an individual. "Life is sustained only by a group of organisms of many species—and their environment . . . Individuals are alive," Botkin maintains, "but an individual cannot sustain life."[70] Life is essentially diverse and interrelated; it depends on change. Rather than being an underlying concept to categorize different beings, different individuals, life is a necessary quality of interrelatedness. Life exists because of the changeability, diversity, and complexity of the relationships between the organisms that exist on the earth and their environments.

When this notion of life is applied to forests, the practice of clear-cutting becomes terribly problematic. The argument that all will be fine in a few generations due to the replanting of trees reduces life to a characteristic of individual species. A forest is not just a stand of trees, but is a complex life-support system, which may have taken centuries to evolve. When life is thought of in terms of entire ecosystems, rather than as a characteristic of an individual, the argument that replanting is a sufficient response to clear-cutting becomes difficult to sustain.

The move away from notions of permanence, fixity, regularity, which is now, perhaps, occurring, removes the simple standard against which Botkin argues we have judged our actions. We can no longer be the shepherds or guardians (or masters) as Marsh would call us to be. Or, at least, the task can no longer be seen in such clear-cut terms. How are we to shepherd a world that is in continual flux? If change is intrinsic to nature, how are we to understand the effects our actions have upon the earth? The ecological problematic of Marsh has been exploded by contingency. We no longer have a nature of permanence to measure our actions against; there is no "original nature" on which to base our decisions.

This new ecological problematic (re)raises the question of management (perhaps to problematize its standing). "How do you manage something that is always changing?"[71] Should forest fires be left uncontrolled? From where do our valuations on change emerge? What changes can we say are "natural"? If we no longer have a solid standpoint, a permanent, unchanging nature from which to judge our effects upon the world, aren't we thrown into a relativistic paralysis? Or worse yet, aren't we turned loose upon the earth, free to do anything and everything? In short, how can we possibly articulate an ethic of ecology in such a world, let alone a politics of ecology?

TOWARD AN ECOLOGICAL ETHIC OF CHAOS

Both Nietzsche and the chaos scientists I presented attack shadowy representations of nature. But the attack is not an end in itself. Struggling with the shadow of God is not a solution, but a problematic. Promise as well as danger exists here. The possibility for environmental degradation exists whether God's shadow lingers on or not. It is not enough simply to attack shadowy conceptions of nature. A new ethic is needed as well.

If, as Botkin claims, the windless pond from which we took our reflection and in which we anchored our place in the world is no more, where can we draw our identity from now? How are we to relate to this chaotic nature? On what are we to base our actions? How will it be possible to act? Is a ground available for a new ecological ethic? Are we not now in a similar position to the one set out by Nietzsche's madman?

Have you not heard of the madman who lit a lantern on a bright morning and ran to the market crying, "I seek God! I seek God!"? Receiving no response other than laughter and jests from those gathered there, he

> jumped into their midst and pierced them with his eyes. "Whither is God?" he cried; "I will tell you. *We have killed him* . . . But how did we do this? . . . What were we doing when we unchained this earth from its sun? Whither is it moving now? Whither are we moving? . . . Backward, sideward, forward, in all directions? Is there still any up or down? . . .
>
> "How shall we comfort ourselves, the murderers of all murderers? . . . What festivals of atonement, what sacred games shall we have to invent? . . . Must we ourselves not become gods simply to appear worthy of it? . . ."[72]

Nietzsche's thought has been widely read as a championing of 'man' becoming God. Nietzsche is typically cast as the hailer of the Promethean Man, the one who would put "man in God's place"[73] to complete 'man''s mastery over the world. His concept of "will to power" has been equated with human domination of nature.[74] Indeed, Nietzsche has been read as the antiecologist; and this reading is, no doubt, powerful.[75]

Such a reading fails, I will argue, to come to terms with certain ecological flows within the thought of Nietzsche to which I have already alluded.[76] Nietzsche's thought, when read with attention to these ecological flows, contests the "man in God's place" position, or the "assuming dominion of the earth" position.

First, there is Nietzsche's perspectivism and his belief that there is no necessity for 'men' to exist, glimpsed earlier, which strongly contest any reading of Nietzsche's philosophy as advocating that 'man' take God's place in the world. Nietzsche's insistence that we all ask the question "What is nature, for me?" is incompatible, as I read it, with assuming dominion of the earth, or taking God's place in the world. It is from this latter position that thinkers insist that their interpretation of the world is true. On my reading, this is fundamentally anti-Nietzschean. "Tyranny," Sarah Kofman contends in *Nietzsche and Metaphor*, "is reprehensible in all its forms, including that of any philosopher seeking to raise spontaneous evaluation to the status of an absolute value."[77] In this sense, Nietzsche's thought is thoroughly antityrannical.

Second, Nietzsche does not hail the murder of God as the ultimate achievement in the evolution of 'man'; Nietzsche longs for the overcoming of even this 'man.' Upon meeting the murderer of God, Zarathustra, in Nietzsche's *Thus Spoke Zarathustra*, exclaims: "I recognize you well, *you are the murderer of God!* . . . You could not *bear* him who saw *you*—who always saw you through and through, you ugliest man! You took revenge upon this witness!"[78] For Deleuze, this man is the reactive man, "[he] puts God to death . . . he wants to be alone with his triumph and strength. *He puts himself in God's place.*"[79] This is not Nietzsche's *Übermensch*. After his confrontation with the murderer of God, "the ugliest man," Zarathustra remarks: "Man, however, is something that must be overcome."[80] It is the ugliest man who kills God, unable to bear the presence of a constant witness to 'his' actions. This 'man' takes God's place.

This 'man' may become an "ecologist," however, assuming responsibility (as God did) for the condition of the earth. Such an "ecological" rendition of Nietzsche's thought can be found in the contemporary German philosopher Hans Blumenberg's reading. It is worthwhile to present Blumenberg's reading of Nietzsche since it follows my own in many important respects, while ultimately diverging on the reading of Nietzsche on nature. Therefore, it may prove beneficial to articulating the particularities of the Nietzschean eco-ethic I am trying to formulate.

Nietzsche formulated the situation of man in the "disappearance of order," abandoned by natural providence and made responsible for himself, but he did so not in order to express disappointment at the loss of the cosmos but rather to celebrate the triumph of man awakened to himself from the cosmic illusion and to assure him of his power over his future.[81]

Blumenberg reads Nietzsche as fostering the conditions through which "the triumph of man" leads to 'man' taking "responsibility for the condition of the world as a challenge relating to the future."[82] In this way, Blumenberg can be read as putting an environmental spin on the celebration of the last man.

Blumenberg accepts, with Nietzsche, that the death of God, or the emergence of modern natural science, does not necessarily entail the disappearance of God's shadow, or "teleological premises."[83] What Nietzsche demonstrates through his critique of modern science, according to Blumenberg, is "how even the great instrument of self-assertion, modern science, stands under a residuum of the conditions whose acceptance in the ancient world and the Middle Ages had kept the will to self-assertion latent."[84] The power of Nietzsche's critique of modern science frees 'man,' according to Blumenberg, from creeping teleologies that ultimately serve only as a burdensome crutch. "How can anyone presume to speak of a destiny of the earth?" Blumenberg quotes Nietzsche as asking. "Mankind must be able to stand without leaning on anything like that."[85]

Freed from such crutches, Blumenberg asserts that 'man' is now capable, through self-assertion, of taking responsibility for the future condition of the world. With the last of the creeping teleologies vanquished by Nietzsche, 'man' stands alone. Blumenberg's "ecology" differs from that of Hallman (glimpsed earlier) most significantly in its rejection of teleology. Hallman requires a telos to still be present in the world in order for 'man' to be able finally to live fully in it. Blumenberg does not seek such a home to (re)turn to. Instead, he celebrates the loss of a teleological home for humanity. The absence of this home is necessary in order for Blumenberg to charge humanity with responsibility for the future condition of the world. It is also necessary for the world to be an open possibility for human action: "'unfinished,' and thus material at man's disposal."[86]

Blumenberg's rejection of the teleology takes an important step, in my mind, away from the shadow of God. Yet his appeal to a world that is unfinished "material at man's disposal" hangs perilously close to cosmologies that present the earth as a gift to human beings to make use of as we see fit. Ultimately, a Blumenbergian ecoethic clothes itself in the apparel of stewardship—albeit a nonteleological stewardship. Human beings are still privileged in Blumenberg's world, perhaps only because we have the capacity to radically alter our environment. And this is an aspect of human beings that must not be brushed aside. But the dangerous aspect of Blumenberg's ecology is that in order for human self-assertion to triumph the world must become plastic.

Blumenberg's reading of Nietzsche and his ecology never consider such statements of Nietzsche as this: "But perhaps this is the most powerful magic of life, it is covered by a veil interwoven with gold, a veil of beautiful possibilities, sparkling with promise, resistance, bashfulness, mockery, pity, and seduction."[87] Nietzsche's vision of life, nature, the world carries with it elements that resist the "material at man's disposal" reading, or the stewardship reading. A nature that is in all eternity chaos further suggests a break from the nature-as-plastic reading Blumenberg presents. If nature is chaos, rather than plastic, it will confound our ability to mold it, to shape it into what we want it to, or think it should be. Moreover, Nietzsche even criticizes our insistences that nature be what we claim it to be, when he argues for not divesting nature of its rich ambiguity, for not attempting to rid nature of what is "nature" in it.

Nietzsche's conception of nature evokes what one might even call a nontheistic reverence for the difference of nature. Blumenberg appears unable to afford nature such respect, for in his world, reverence is due to a God, or to no thing at all. Thus a world stripped of its teleological bindings becomes a plastic world. But God is dead, and Nietzsche, who would vanquish God's shadow, too, still speaks of having a reverence to nature. Nature receives Nietzsche's reverence not from being predisposed to us, not from being uniform, regular, or even unfinished material waiting for us to finish it. Nietzsche's reverence toward nature emerges from its ambiguous possibilities, its contingency, its resistance as well as its promise, its mockery, its bashfulness, and its ability to confound human self-assertion.

The rejection of openly, or shadowy, theological underpinnings for nature and our relationship to it need not release humans to assert their mastery over the world, even if that would involve taking responsibility for the future condition of the world. Rather than cutting us loose to do anything at all, a Nietzschean eco-ethic inserts a profound and troubling element of caution, that in our actions we may be imposing our designs and interpretations on the world and the vast diversity of life that occupies it along with us.

The caution that emerges from Nietzsche's reading of nature and human beings' relationship to it contains several important elements that distinguish it from the caution called for by Marsh. I will take the time here to highlight one. The caution Marsh spoke of arose from a belief that 'man' had taken a wrong turn and was no longer interacting with the world that was created for him in the appropriate way. Marsh's ecology, despite its radicalness in the nineteenth century, is still linked to a Lockean notion of use and property, discussed in the Introduction. The only place a Marsh and a Locke differ is on the extent to which humans were divinely directed to use the earth. Where Locke's conception of property and use of nature accompanies a belief that God had provided for us, Marsh's ecology emerges from a recognition that God did not provide for us regardless of our actions. But, rather than emerging from the shadow of God, Marsh fell back on a belief that indeed the world was made for us, only we have forgotten the role we were supposed to play.

Nietzsche attempts to break from this human conceit completely and articulate a new ethics—a nontheological ethics. For Nietzsche, we are under no divine orders to subdue or care for the earth. The death of God and the vanquishing of God's shadow need not necessarily lead to the technological domination of the world by human beings. We are fast accomplishing this from within God's shadow. From Nietzsche's perspective, the earth in no way belongs to us. It was not placed here for us to use. We are, instead, only one of billions of all-too-brief occupants of an eternally chaotic world. Perhaps we ought to take that thought seriously.

NIETZSCHE, ECO-ETHICS, AND THE STATE

A final element in Nietzsche's potential eco-ethics involves the question of where. A funny question to raise, perhaps, when thinking about ethics. Although numerous political theorists from Aristotle to Hegel have placed ethics within the boundaries of the state in the sense that ethics provide a foundation for the state, Nietzsche's critique of the state suggests a different space for his ethics. Speaking of the state as a foul-smelling, cold monster and a new idol, Zarathustra urges us to "break the windows and leap to freedom . . . only where the state ends, there begins the human being who is not superfluous: there begins the song of necessity . . . Where the state *ends* . . . Do you not see it, the rainbow and the bridges of the overman?"[88]

Nietzsche attacks the state for a variety of reasons, from its lies about being "the people," to the way it sacrifices humans for its purposes, to its lies about being the most important creature on earth. In order to overcome these, and other, aspects of the state, Nietzsche suggests looking to the end of the state. And here I read him not to be making a temporal suggestion but a geographic one. The overman, that human who exists beyond resentment against the nature of things, that human who refuses to divest existence of its rich ambiguity, is not a creature of the state, if, as I suggested through a brief glimpse at Locke's theory of property in my introduction, the state can be read as the product of both a particular orientation to nature and a particular type of human. Rather than being a creature who resides in the space of sovereign territory, the Nietzschean overman may best be said to exist where the state ends. Following a number of post-Nietzscheans, whom I will engage in the next three chapters, where the state ends is a multidirectional and polymorphous space.

Moreover, Nietzsche's perspectivism, his radical reading of interpretation, throws into question the concept of space at work in discourses of state sovereignty. Zarathustra speaks of the lies that spew from the mouth of that "new idol" the state. "On earth there is nothing greater than I: the ordering finger of God am I."[89] The problem, Zarathustra contends, is that people believe the state. They accept its statement concerning the space of politics, among many other things, as though it were carved into stone tablets. But it is only one interpretation.

2

Exploring the Space of the
(Inter)State (I): Sovereignty

In a word, the origin and history of the concept of sovereignty are
closely linked with the nature, the origin and the history of the state.

F. H. HINSLEY 1966

If we are interested in the space of ecopolitics, surely we are interested in the space of the state. Since at least Aristotle, politics has
been thought of as existing primarily within the confines of, in very
general terms, the state. Moreover, and more recently, politics and
the state have become fused with a concept of sovereignty. Sovereignty historically has been associated with a rule of law, involving a
person or a group of people with the authority to establish laws over
a given geographic area and a given people. In modern political-
science terminology, only a state can be thought of as displaying sovereignty. But does ecopolitics operate within the space of the state,
within the space of sovereignty?

On the first and perhaps most obvious level, ecopolitics challenges the boundaries of sovereign territory as it slices effortlessly
from one state to another, in the form of acid rain, chlorofluoro-
carbons (CFCs), radioactive fallout, polluted waterways, and so on.
From this perspective, ecopolitics might more properly be said to
occupy the space of international political theory, for international
political theory is said to involve problems that exceed separate sovereign territorial state boundaries. Indeed, there is a growing amount

of literature about ecopolitics from an international political theory perspective.[1]

Because of the need to distinguish international politics from state politics, international political theorists have had to distinguish the space of the state from the space of the interstate. This distinction is not solely geographic. The construction of political space is not limited to drawing a boundary around a tract of land and setting it apart from other tracts of land. The construction of political space involves other types of boundaries as well: boundaries that exclude certain institutions or issues from the realm of the political; boundaries that exclude certain communities from political discourse by defining them as outside the realm of politics—the state; boundaries that exclude certain types of analyses by rigidly setting the terms of political discourse. All of these boundaries help to construct the space of politics, whether it be state or interstate.

Put bluntly, the map of international politics, as it is generally represented in the field of international political theory, consists of sovereign states in a condition of global anarchy. Although numerous international organizations such as NATO, the European Union, and the United Nations can exist, the principal characteristic of the interstate is the absence of a unifying authority (sovereignty) across the globe that can effectively organize the many separate sovereign states into one cohesive body. International political theory, traditionally, is about how these separate sovereign states relate to one another within this anarchic condition. But it is much more than that. It is also very much about theorizing sovereign state territory. Across the field of international political thought, from realists to idealists, the anchor of international political theory can be found in the space of the sovereign territorial state.

In exploring the space of the (inter)state through international political theory, I will focus on two theorists said to occupy somewhat opposing positions within this discipline: Kenneth Waltz and Hedley Bull. Hailed as an originator of "neorealism," Waltz's contribution to the field of international politics has been claimed to amount to nothing less than the "systematiz[ing of] political realism into a rigorous, deductive systemic theory of international politics."[2] Where political realists had effectively delineated some of the operative assumptions about the character of politics in general and the relations between states in particular, according to international relations theorist Robert Keohane they were "unable to create a consistent and convincing theory."[3] Waltz's theoretical contribution to the field helped enable it to move beyond "reliance on the nature of human beings to account for discord and cooperation in world politics, [and] focus instead on the competitive, anarchic nature of world politics as a whole."[4] In short, Waltz focused the theory of international politics onto the problem of sovereign, territorial states existing in an anarchic realm. In other words, Waltz focused attention on the structure of international politics.

FINDING THE STRUCTURE OF THE (INTER)STATE

> When we describe structure . . . we are in the realm of grammar and syntax, not in the realm of the spoken word. We discern structure in the "concrete reality" of social events only by virtue of having first established structure by abstraction from "concrete reality."[5]

"Structure is not something we see," Waltz claims.[6] We "find" it through a process of abstraction from "concrete reality." In other words, the structure of the international arena is not "really" there. It is placed there by theorists of international relations who then set out to explain the actions of international actors by the presence of this structure. There is a self-fulfilling prophecy to this type of theorizing that recalls the discussion in chapter 1, regarding the ecologist's equations: "No good ecologist ever forgot that his equations were vastly oversimplified versions of real phenomena. The whole point of oversimplifying was to model regularity. Why go to all that trouble just to see chaos?"[7] As with the "good ecologists" who are intent on modeling regularity, Waltz is intent on finding structure. He is aware that his theorizing oversimplifies the complexity of the global political arena. But basing his theory on "concrete reality" will not allow Waltz to "see" the structure that will provide his theory with sufficient explanatory power:

> In reality, everything is related to everything else, and one domain cannot be separated from others. Theory isolates one realm from all others in order to deal with it intellectually. To isolate a realm is a precondition to developing a theory that will explain what goes on within it . . . The question, as ever with theories, is not whether the isolation of a realm is realistic, but whether it is useful.[8]

Waltz's claim about interrelatedness could be seen as an opening for an ecological analysis of international politics—interrelatedness being one of the principal elements of ecology. A theory of international politics based on interrelatedness and the inability to separate one domain from another might follow directly from the Nietzschean eco-ethic presented in chapter 1, drawing on the rich ambiguity of existence and refusing to divest existence of this ambiguity. It might cause an international relations theorist to take seriously the multiple ways in which (eco)politics slices across pathetically porous sovereign state boundaries. It might cause an international relations theorist to pause before positing the sovereign state as an unproblematically unified site of politics. But Waltz's theory is not about interrelatedness; it is not about, to use his term, reality.

Instead, Waltz constructs a theory of structure, a theory of intellectual use—interrelatedness apparently being an intellectually unusable concept. A useful theory, Waltz instructs us, involves the isolation of one realm from another. In other words, political territories must be (theoretically) isolated from one another if a theory of (international) politics is to have any use.

There is a tension between the first sentence of Waltz's passage, just cited, and the last: between "reality" and "useful"; between interrelatedness being essential to life and the isolation of realms being essential for use; between conditions necessary for life and conditions necessary for knowledge. Life requires difference; knowledge requires similarity.[9] The latter attempts to encapsulate the former, but never succeeds. Deleuze, in his reading of Nietzsche, comments that life is "opposed to" knowledge, in that "life goes beyond the limits that knowledge fixes for it."[10] Privileging one element in this tension over the other will not eliminate the tension. To focus solely on the conditions of life will not free us of the problems of knowledge. To emphasize knowledge—or, to use Waltz's terms, explanatory or useful theory—will not eliminate the excessiveness and ambiguity of reality. In order to articulate an ecological political theory, following Nietzsche, one must accept the necessary tension between knowledge and life, or "reality" and "useful."

Waltz's theory of international politics, however, attempts to eliminate this tension through a privileging of explanation or use. "Reality" becomes an obstacle to formulating a "theory that will explain." If we are to explain what goes on within a certain realm, Waltz insists, we must isolate that realm. Interrelatedness overcomplicates the matter. Isolating a realm is not, on my reading, a problem in and of itself, if the theorist continues to foreground the inter-relatedness of reality.[11]

Waltz asserts that what is important for theories is not whether they are "realistic"—that is, whether they come to terms with the interrelatedness of reality—but whether they are useful. Or, as he puts it elsewhere, "the problem is to develop theoretically useful concepts to replace" vague and varying notions "such as environment, situation, context, and milieu."[12] "Useful" concepts must replace "realistic" ones. It is not just that the realms of reality must be isolated, separated, but that the interrelatedness of reality must be abstracted away in order to develop a theory that will be useful.

"Clear and fixed meaning" must be given to "reality." This meaning is given to reality in Waltz's texts through processes of abstraction. What Waltz wants to see is structure; and structure, he has informed us, can be found in reality only by first placing it there through a process of abstraction (one comes to wonder how such theorizing can possibly be called [even neo]realist).

Driven by the need to see a structure in the world, Waltz constructs a theoretical framework that allows him to see a "continuity," "consistency," "recurrence," "repetition," "persistence," "enduring character," "striking sameness" to international politics.[13] By moving further "away from everyday reality and by . . . lower(ing) 'the degree of empiricism involved in solving problems,'"[14] Waltz is able to re-present a world that fits neatly within his structural framework: a world of bounded interacting units (sovereign states) in an anarchic realm. The "reality" of the world is successfully glossed over within Waltz's structural representation of it. He is able to marginalize, or even discount, those who would construct theories in "nonstructural" ways.

Waltz does just that in an essay written for Keohane's *Neorealism and Its Critics*. Taking up the criticisms of Richard Ashley from "The Poverty of Neorealism" (reprinted in that collection), he claims that reading Ashley is "like entering a maze," that Ashley "shift[s] from one view to another."[15] Ashley is involved in the project of critical theory, according to Waltz, which "seeks to interpret the world historically and philosophically." Waltz, on the other hand, is engaged in problem-solving theory, dedicated to "understand[ing] and explain[ing the world]."[16] Ultimately, Waltz concludes: "Ashley's critical essay reveals to me no clue about how to write an improved theory of the latter sort."[17] It is in this move that Ashley's project is marginalized. The attacks that Ashley levels at Waltz's necessary assumptions are discarded as unhelpful, since they do not improve the effectiveness of problem-solving theory; instead, they highlight the interrelatedness of reality, perhaps making theory more "realistic," but less "useful."

DRAWING THE BOUNDARIES OF THE (INTER)STATE

In considering the structure of international politics as set out by Waltz, it is important to pay attention to the lines within his text, to become intimate with these boundaries. How have they been drawn? Where have they been drawn? How are *they* used? These questions, it seems to me, ought to guide one's reading of Waltz's texts. The units of Waltz's system created by the drawing of these lines must be considered to be as important as the system itself. What lies within the border lines of the interacting units, as well as what is closed out by them, must be considered. In short, an effort must be made to read Waltz's theory of international politics contrary to the orders he gives for its reading: to read it so that the ecology of the text and the ecology of (international) politics may speak; to read it so that the interrelatedness, contingency, ambiguity, and diversity of text and globe are not drowned out by the utility of his structural theory.

Waltz's two main texts, *Man, the State and War* and *Theory of International Politics,* argue for a structural approach to international politics against reductionist, unit, statist, or psychological approaches. As in the latter, the former argues that the method to be employed is "proceed by examining assumptions and asking repeatedly what differences they make."[18] International political theory is the realm of assumptions. We cannot, we are told, attempt to explain reality by describing it. We must step back from the object; we must make abstractions, make assumptions, delineate, differentiate, draw boundary lines. Waltz informs us how theory in general and international political theory in particular must be constructed: "first, one must conceive of international politics as a bounded realm or domain."[19] Not only are lines to be drawn within the realm, separating the various elements within it, but first a line must be drawn about the field as a whole—to set off the domain of international politics from all other fields of study. What is this domain, this bounded realm of international politics?

If international politics is understood primarily by its anarchic character—that is, a lack of order—how can a theory of international politics be constructed? If the first rule for developing a theory is to conceive of a bounded realm or domain, the anarchy of international politics seems to suggest that a theory of international politics is not possible. Waltz recognizes this problem:

> Structure is an organizational concept. The prominent characteristic of international politics, however, seems to be the lack of order and of organization. How can one think of international politics as being any kind of an order at all? . . . In looking for international structure, one is brought face to face with the invisible, an uncomfortable position to be in.[20]

In an attempt to overcome this problem of looking for what appears not to be there, Waltz draws an analogy between the international political system and the free-market economic system. "Reasoning by analogy is helpful," Waltz maintains, "where one can move from a domain for which theory is well developed to one where it is not. Reasoning by analogy is permissible where different domains are structurally similar."[21]

Although I will attempt to accept his analogy in order to get on with his argument—the building of his structure—I would like to raise some questions. It is not clear to me that even classical free-market economics is as anarchic a space as Waltz wants us to believe international politics is. Perhaps there are no government controls by way of price-fixing, taxes, distribution regulations, and so on in a classical free-market economic system, but can Waltz make the argument that there is no hierarchy to this system? Without the force of a government, or even the existence of some form of society or community, some form of laws, norms, and so on, could the economic units even begin to carry out their business? Isn't there a necessity for society to be, for the most part, peaceful and ordered—that is, must it not at least respect laws of property? This is similar to the point that Bull makes concerning the society of states in his *The Anarchical Society*: that states in the international arena share a system of values that allows for the business of international relations to take place.[22]

Waltz, however, avoids the appeal to values in favor of focusing solely on the structural characteristics of systems. Through his appeal to the free-market economic arena Waltz reaches the conclusion that "international-political systems, like economic systems, are formed by the coaction of self-regarding units. International structures are defined in terms of the primary political units of an era, be they city states, empires, or nations."[23] For Waltz, the primary aspect of both of these systems is not values but actions. What determines the structure of each system is the action of the units in question: in the economic system, the behavior of firms, and in the international system, that of sovereign states. Sovereign states emerge as the anchors for Waltz's theory of international politics. With no overarching order available to establish a theory, Waltz finds the definitive foundation of international politics in states-as-actors.

The border line of the state is then quickly drawn. "Domestic systems are centralized and hierarchic."[24] The presence of centralization and hierarchy in these units allows Waltz to enter the anarchy of the international arena. Any ambiguity within the sovereign state must be abstracted away if ground for a structural theory of international politics is to be found.

It is in the presence of this grounding that Waltz makes his case for how ecological problems, even global ones, must be handled. Simply put, Waltz declares that despite problems being found at the global level, "solutions to the problems continue to depend on national policies."[25] His construction of the state and interstate system guarantees that this remains the only means available. The border lines that frame sovereign territorial states, the primary international actors, according to Waltz, successfully sequester politics within state spheres. The potential to solve global problems exists only inside these borders, for outside, anarchy reigns; and inside, there is only this hierarchical and centralized space. Waltz's response to the interrelatedness, the excessiveness, the ambiguity of ecological problems is intended to confound globalists (the only adversaries Waltz can see) who claim that global problems require global solutions by a global body.[26] Such an appeal, Waltz insists, ignores the structure of the international realm. Wishing for an international authority to deal with international problems does not make it so. States exist within a realm of anarchy, a self-help system; for Waltz the answer to global ecological problems is obvious. Solutions to the problems continue to depend on nation-states in general and the major powers in particular.[27]

The state as a sovereign, politically autonomous, bounded, self-regarding, acting unit is "given." It is summoned as an original (re)-source. Its originality is reflected in the "enduring anarchic character of international politics. . . the striking sameness in the quality of international life through the millennia."[28] Moreover, the original location of the state is further supported by the "wide assent" that international political theorists have given to this view of international life across the centuries.[29] The anarchic system of international politics requires the (enduring) presence of sovereign territorial units to give it its structural character; similarly, these units must have an anarchic interstate to establish their internal structure and external boundaries:

> National politics is the realm of authority, of administration, and of law. International politics is the realm of power, of struggle, and of accommodation. The international realm is preeminently a political one. The national realm is variously described as being hierarchic, vertical, centralized, heterogeneous, directed, and contrived; the international realm, as being anarchic, horizontal, decentralized, homogeneous, undirected, and mutually adaptive.[30]

Having inscribed international space by locating the sovereign state within it, Waltz draws the border of the state by opposing it to the interstate—that which required the state to already be there to give it its character. The structure, for all intents and purposes, is in the place. All that is needed is all that is there—states in anarchy. Or, as Waltz puts it: "the structure of the system and its interacting units."[31]

Two moves are being made here. While international politics is being established as a decentralized realm grounded in the structure of the various sovereign state entities that populate it, national politics is established as a centralized realm in order to provide the necessarily hierarchical spaces to give the anarchical space of international politics its structure. The state must be an unproblematic unified site for Waltz; otherwise his theory of international politics has no foundation—and Waltz cannot envision a theory (a structural theory, at least) without a firm foundation.

A major portion of my argument in this work is that this conception of sovereign state politics is problematic. To construct an unambiguous relation between sovereignty and territory eliminates the interrelatedness of political reality that exists within and between these geographic boundaries and the rich ambiguity of political existence that swirls about inside the territorial boundaries of the sovereign state. The possibility for a Nietzschean eco-ethic is delivered a death-blow through this move. The presence of hierarchical, vertical, cen-tralized, directed territories known as sovereign states squeezes out the space(s) in which my Nietzschean eco-ethic operates—not just within the state but across the globe as well, due to the structure that encompasses the global arena. How can an eco-ethic of difference find space to operate in a world structured by the presence of Waltz's sovereign territorial states?

STATES AND "NOT-STATES"

If we venture across the field of international political theory, from the neorealist Waltz to the neo-idealist Hedley Bull, we can find re-markable similarities in their constructions of state space, despite their apparent differences.[32] Where Waltz is forced to locate the structure of the international system solely in the acts of the individ-ual units of the system, Bull maintains that what holds the interna-tional system together is the culture of "modernity."[33]

What Bull means by this is that, unlike the realist, Hobbesian, or Waltzian position that maintains that states exist in a condition of absolute anarchy or a war of all against all, states' actions are in-stead tempered by "common rules and institutions" that are exem-plary of the modern age—that period extending (in the West) from around the sixteenth century to the present. Virtually all states in the late twentieth century, Bull proposes, are infused with a common set of values that emerged in Europe some three or four centuries ago. Where Waltz's position suggests that the only thing that brings states together is fear of power, or self-interest, Bull argues that the actors known as states exist in an anarchical *society* bound by shared polit-ical ideals.

The question of what a state is must still be addressed. For Waltz, states are the centralized, hierarchic units that form the structure of the international political system. Bull's definition is, in some ways, more straightforward. All states, he writes, "possess a government and assert sovereignty in relation to a particular portion of the earth's surface and a particular segment of the human population."[34] Government, sovereignty, territory, population: Bull establishes the four pillars of the modern state. In this regard, he is able to differen-tiate the political community he calls the state from other political communities that may not have the support of all four of these pillars. Bull's attempt to solidify the character of the state, then, provides a glimpse into a political space characterized by the "not-state."

A community that "merely claims a right to sovereignty," for example, but is unable to assert this right in practice is not a state, according to Bull.[35] If one group within a particular population is unable to solidify its claim to sovereignty against other groups within the same population, no state can exist. Similarly, if a population is unable to quell challenges to its sovereignty from "outside," from other states perhaps, no state can exist either. Maintaining control over a population is a necessary but not a sufficient condition for being a state, however. A community in which a ruler asserts (a recognized) supremacy over a people but not over a distinct territory can also not be said to be a state.[36] States require territory. The Roman Catholic Church, for example, does not fit the description of state being used here. Furthermore, political communities, such as those "in parts of Africa, Australia and Oceania, before the European intrusion . . . [which] were independent political communities held together by ties of lineage or kinship," were (are) not states, since "there was no such institution as government."[37] A political community that encompasses a specific territory and involves a particular population still cannot be considered a state, on Bull's account, if it does not contain the proper governmental institutions. In order to be a state, according to his analysis, population, sovereignty, territory, and government all must be present in the community.

By setting off particular political communities as "not-states," Bull is able to exclude these forms from further analysis. He is also able, as is Waltz, to establish both the solid space of the sovereign state and hence the domain of international politics. "The relations of these [other] political communities might be encompassed in a wider theory of relations of *powers* . . . but [they] lie outside the domain of 'international relations' in the strict sense."[38] International politics is about states. These other political communities, not being states, are not within the international political theorists' field of study since they do not contain the necessary foursome of population, sovereignty, territory, and government. But Bull utilizes rather limited conceptions of each.

It is important to recognize in Bull's conceptualization of the state not just that a division has been constructed, but how it has been constructed. What has been excluded in order to articulate this concept of the state? Why have these communities been excluded? What notions of territory, government, population are excluded from Bull's analysis? In what ways is political thought limited by these exclusions? How might it differ if it were extended into these not-state communities?

SILENCING SOVEREIGNTY

Despite their different readings of the international system of states, Bull's and Waltz's portrayals of the state are all too similar. Both fit into a long line of Western political theory, dating back to the sixteenth or the seventeenth century, that links the state (and ultimately, politics) to the space of sovereign territory. There has been, global political theorist Rob Walker argues, a

> relative silence about sovereignty in Western political theory . . . a silence that marks the acceptance of the rearticulation of political space, the shared sense of a sovereignty within and an anarchy without, and the displacement of thinking about sovereignty into a concern with who may rightfully claim sovereignty within a particular state.[39]

Reading the histories of state and sovereignty back through this silence allows theorists to find "a striking sameness . . . throughout the millennia"[40] in the practice of (international) politics. Thucydides' *History of the Peloponnesian War*, written in the fifth century B.C.E., becomes a necessary text for students of international relations some twenty-five centuries later, as though the events described by Thucydides mirror events of today.[41] Machiavelli's early sixteenth-century work *The Prince* gets situated into the continuous "tradition" of international relations texts with little or no discussion of Machiavelli's theoretical challenges to the political theory of his day, specifically with respect to conceptions of sovereignty, territory, and the state.[42]

International politics, we have been taught, is about solid sovereign entities known as states—it has always been so and will always be so. As long as there have been states, there has been sovereignty and vice versa, runs the syllogism of international political theory. But to assume this connection between state and sovereignty is to make an assumption that the only stable form of government is a state that can make and enforce its own policies, and contain politics within its own bounds . . . an assumption . . . like the late medieval political theorists who thought the choice was between the papacy and the empire as the primary unit of political organization. The best of these theorists realized that something new was emerging—the modern state; but they lacked the categories for understanding this new reality. We now are in a similar position.[43]

Our question no longer involves thinking politics outside the confines of the papacy or the empire, but rather thinking politics outside the confines of the sovereign territorial state. Why must we, theorists of politics, remain confined within territorially bounded conceptions of the political? Perhaps it is time (yet again) to rethink our spatial assumptions of politics, to look beyond the border of the sovereign territorial state and its relationship with other sovereign territorial states, to think politics differently, in a different space—a not-state space?

To think in the space of the not-state is not as easy as it might appear. Thought about politics, as Walker has suggested, has been structured over the centuries by the model of the sovereign territorial state. Moreover, contemporary political philosophers Gilles Deleuze and Félix Guattari claim, "thought as such is already in conformity with a model that it borrows from the State apparatus and which defines for it goals and paths, conduits, channels, organs, an entire *organon*."[44] It is not just thoughts about politics, but thoughts in general, Deleuze and Guattari argue, that have been structured by the state—that is, sovereignty. What this entails will, it is hoped, become clearer as I proceed. For the moment, let me suggest that what Deleuze and Guattari's theorizations can offer is, perhaps, a "chaos theory" for political thought, a political theory in which a Nietzschean eco-ethic can survive. As with the chaos theorists examined in chapter 1, Deleuze and Guattari attack a model of thought that depends on a representation of the world as highly structured, regulated, ordered. For Deleuze and Guattari, such thought is state thought, tree thought, binary thought.

If we are to explore a political space that is not highly structured, regulated, ordered, consistent, repetitious, and so on, then perhaps we need "to extricate thought from the State model."[45] The politics of ecology exceeds the limits of the state; and we need to allow theory to exceed the state model as well. If state thought mirrors an image of a tree, according to Deleuze and Guattari, then not-state thought may be said to follow rhizomatic lines.

RHIZOMES AND TREES

Deleuze and Guattari's attempt to extricate thought from the state model maps political landscape differently. Their map is not centered around a sovereign authority. It does not represent a territory shaped by a sovereign presence. Such a territory is, in the minds of Deleuze and Guattari, arboreal—treelike. It has roots that sink deep into the ground, fixing a particular place. It has a single trunk that branches out, sending aspects of itself away from the center while always remaining connected to a unifying trunk, that in turn is firmly rooted in one place. The map that Deleuze and Guattari construct is far more rhizomatic than arboreal.

A rhizome is a system of roots and shoots that grow along or under the ground, extending in many different directions, reaching into many different spaces. Horizontal, not vertical. Waltz differentiates state space from interstate space by arguing that the former is vertical, whereas the latter is horizontal. But, as we shall see, Waltz's description of international politics is far from Deleuze and Guattari's description of the rhizomatic. Where Waltz wants to bypass interrelatedness in constructing his structure of international politics, Deleuze and Guattari throw themselves headlong into the chaos of a reality where everything is related to everything else. "Principles of connection and heterogeneity: any point of a rhizome can be connected to anything other, and must be."[46] States in the international realm, as presented by Waltz or Bull, do not constitute a rhizome but perhaps a forest, where numerous individual trees compete for space and light and water. Deleuze and Guattari's attempt to articulate a rhizomatic politics involves an entry into political space far different from that thought by Waltz or Bull.

Deleuze and Guattari offer the rhizome as one metaphor for their theory, not in opposition to the root-tree model, which they claim is dominant within Western thought, but as different from it. The tree occupies a special place in Western thought, Deleuze and Guattari contend. The tree plots a point, it fixes an order.[47] The tree model cuts off possibilities. It searches for ones, not manys. "The West has a special relation to the forest," Deleuze and Guattari assert, "and

[to] deforestation; the fields carved from the forest are populated with seed plants produced by cultivation based on species lineage of the arborescent type."[48] Western thought and action are based on monocultures. Through their reification of the sovereign territorial state, Waltz and Bull attempt to produce a monoculture politics. The sovereign territorial state becomes *the* political unit.

"The rhizome is altogether different, a *map and not a tracing.*"[49] The rhizome is a map on Deleuze and Guattari's reading because it is experimental; it is in contact with the real. A tracing reproduces what is already closed within the unconscious. Waltz's method of theorizing is a tracing, in the way in which it "finds" structure in the world by first establishing it there through a process of abstraction, a process of distancing oneself from reality. A map fosters connections between fields; it is

> open and connectable in all of its dimensions; it is detachable, reversible, susceptible to constant modification. It can be torn, reversed, adapted to any kind of mounting, reworked by an individual, group, or social formation . . . A map has multiple entryways, as opposed to the tracing, which always comes back "to the same."[50]

Where state thought traces territories around a sovereign point of view, rhizomatic thoughts, not-state thoughts, draw their maps "with a more multiple, lateral, and circular ramification."[51] Rhizomatic thoughts are not centered around a specific point of view, but follow flows and think territories in different ways.

If space is (re)thought in this manner, Deleuze and Guattari contend, "the meaning of the Earth completely changes . . . With the legal model [the tree model], one is constantly reterritorializing around a point of view, on a domain, according to a set of constant relations; but with the ambulant model [the rhizomatic model], the process of *de*territorialization constitutes and extends the territory itself."[52] An example of this deterritorialization could be pollutants that have been discharged into the air. The territory of the state is firmly established and organized around a central sovereign point of view. But these airborne particles, which emanate from a particular sovereign territory, flow across state boundaries, thus deterritorializing the sovereign space of the state, yet extending territory on their travels. In other words, "the physical effects of our decisions spill across national frontiers."[53] As long as politics is thought in a static, treelike manner, theorists of global ecopolitics will continue to fail to come to terms with this rhizomatic character of their subject.

To think politics along rhizomatic lines would thoroughly confuse many of the central concepts of political theory. Territory, for instance, would no longer be such a stable object upon which sovereign states could be built. Territories, on a rhizomatic reading, are constantly shifting, contingent, ambiguous, and so on. Think of the plant that deposits a seed into a nearby stream. The territory of this plant is extended by the flow of the stream. Yet this territory is not subsumed by the plant; open spaces exist within this reterritorialized space. In this sense it differs from an arboreal expansion of territory that extends out from a sovereign center, filling more fully the space around it. A rhizomatic extension resists the fixity and solidity assigned to territory by the tree model. Rhizomatic territory is not overcoded in the manner in which arboreal territory is.

What Waltz and Bull do first in constructing their theories of international politics is fix, code, striate a territory that will encompass the state. But, to repeat, rhizomatic territory refuses this fixity; it is fluid. Again, what I think Deleuze and Guattari are offering is a "chaos theory" conception of space, a fractal rather than a Euclidean conception of space. This change in the conception of space and hence the meaning of territory is not the answer, however. Thinking territory in this manner will not (necessarily) solve ecopolitical crises.

"The rhizome includes the best and the worst: potato and crabgrass."[54] Deleuze and Guattari do not offer their rhizomatic model as *the* answer for thought in general, or even political thought in particular. Rhizomatic thought is not "the good" to state thought's "evil"; it is, however, different. And it is a difference that must be engaged by ecopolitical thinkers and actors. While Deleuze and Guattari think that the difference of rhizomatic thought offers positive possibilities, they are not suggesting that this mode of thought will guarantee salvation, that if politics were just conducted on a rhizomatic plan(e) everything would be fine. Indeed, they already believe that much of what is political exists on such a plane.

Late-modern capital, for instance, occupies a rhizomatic, rather than an arboreal, space. Financial flows, the *Financial Times* reported in 1987, are "rapidly becoming indifferent to the constraints of time, place, and currency."[55] They are not oriented around a particular point, distributed according to a set of constant relations; rather, they extend themselves in multiple directions, reconstituting their relations as they go. Global capital is not bound by sovereign territories. It does not orient itself around a particular point on a map. It does not take hold of one area and branch out from there. It follows more rhizomatic lines of satellite communications and trade routes. It has no central root system that fixes it to one place. Where is General Motors rooted? Detroit? The United States?

EXPLORING RHIZOMATIC SPACE

Capital may not be rooted to a specific territory, but it is not completely detached from the world of sovereignty. By necessity, it exists within the lines of sovereign territory while exceeding these lines in multiple directions, thus demonstrating the limits of sovereignty to fix, code, regulate, control its territory. "The State is sovereignty," Deleuze and Guattari assert, "but sovereignty only reigns over what it is capable of internalizing, of appropriating locally."[56] Sovereignty does not regulate everything that exists within its territory. There is a beyond, an outside, that exists within the geopolitical walls of the sovereign state. Deleuze and Guattari are interested in that which sovereignty has not, does not, or cannot internalize. They are interested in both that which is perhaps inside the geopolitical boundaries of the state but *outside* its apparatus and that which cuts across the geopolitical boundaries of the state, thus existing *outside* its territory as well.

Their interest in that which sovereignty cannot appropriate or internalize should not be taken to indicate an interest in the Waltz/Bull-ian realm of international politics, as I have already suggested with respect to their concept of the rhizome. Deleuze and Guattari are not international political theorists. No, "the outside of States," they contend,

> cannot be reduced to "foreign policy," that is, to a set of relations among States . . . The outside appears simultaneously in two directions: huge worldwide machines branched out over the entire *ecumenon* at a given moment, which enjoy a large measure of autonomy in relation to the States; but also the local mechanisms of bands, margins, minorities which continue to affirm the rights of segmentary societies in opposition to the organs of State power.[57]

Deleuze and Guattari, in pursuing the outside of the state apparatus, are not pursuing relations among states. They are challenging traditional divisions within political thought, in particular, a division between a domestic policy inside and a foreign policy outside.

Foreign policy, or international relations, is not outside, from Deleuze and Guattari's perspective. As we have seen through glimpses of Waltz's and Bull's theories of international politics, international politics is very much about what is inside—the construction of sovereign state space and the interactions of these sovereign territorial states. From Deleuze and Guattari's perspective, the outside involves the not-state (which may exist within, without, or across sovereign state boundaries): Exxon, General Motors, the Catholic church, global nongovernmental organizations, local grassroots political movements, indigenous peoples, and so on. For Bull, a thinker of the state, these not-state "communities" fall outside the realm of analysis. For Deleuze and Guattari, they offer an opportunity to think politics differently: to come to terms with the manner in which the outside is always, already inside, and vice versa.

By attempting to incorporate different actors and different spaces into political analysis, Deleuze and Guattari are not saying that states have become insignificant. The striated space of the state has not been replaced by the smooth space of the not-state in their thought. Deleuze and Guattari are interested in the relationship of these spaces, their interdependence, their mix:

> No sooner do we note a simple opposition between the two kinds of space than we must indicate a much more complex difference by virtue of which the successive terms of the oppositions fail to coincide entirely . . . *the two spaces in fact exist only in mixture*; smooth space is constantly being translated, transversed into a striated space; striated space is constantly being reversed, returned to a smooth space.[58]

One might say, using the terminology of Deleuze and Guattari, that smooth space exists between striated space, not in the sense of being bound on all sides by striated space geographically or temporally, but by means of its interaction with striated space. "Being between . . . means that smooth space is controlled by these two flanks, which limit it, oppose its development, and assign it as much as possible a communicational role."[59] But, being between also "means that (smooth space) turns against (its two flanks) . . . affirming . . . a force of *divergence* . . ."[60] The smooth space of global capital is limited, opposed in many ways by the striated space of the state. Tariffs, environmental regulations, occupational safety laws, and so on, all are attempts by states to control capital. Yet the fluidity of capital often subverts attempts at striation by the state. Despite the presence of environmental regulations in one state, capital can often easily slide across territorial borders into the next state, where these regulations may not be in place. However, the resulting pollution may well float back into the territory of the first state, confounding further its attempts at controlling its territory and capital.

The question of indigenous peoples' "sovereignty" or "rights" is also illustrative of Deleuze and Guattari's point. Depending on the policy of the particular state, indigenous peoples may hold a dual status as both citizen and foreigner. The presence of an "Indian Nation" in the midst of the geographic boundaries of a sovereign territorial nation-state should appear troubling, according to Vine Deloria Jr. and Clifford Lytle.[61] How is it that one nation can exist within another? To the U.S. citizen, Deloria and Lytle muse, there is an absurdity to referring to "Indian Nations." Part of that absurdity lies in the assumption that the inside is unambiguous. But the presence of an "Indian Nation" within the sovereign borders of the United States ought to explode the assumption of the unambiguous sovereign inside. It should raise questions concerning the nature of sovereign territory and the concept of space that accompanies it. Although Deleuze and Guattari do not speak directly to the issue of indigenous peoples, their metaphorical use of the nomad suggests a possible line of inquiry.

A NOMADIC ORIENTATION

In his nineteenth-century examination of democracy in America, Alexis de Tocqueville discussed the condition of the North American territory and the relation of the "Indians" to it: "Although the vast country that I have been describing was inhabited by many indigenous tribes, it may justly be said, at the time of its discovery by Europeans, to have formed one great desert"[62]—smooth space. On de Tocqueville's reading (in Deleuze and Guattari's language), the "Indians" distributed themselves in an open space without reterritorializing it around a specific point, without striating it. In order to be civilized, de Tocqueville continued, in order to constitute a nation-state properly speaking, the "wandering propensities" of the "Indians" had to be checked; they had to be "settled permanently."[63]

Is the nomad to be understood solely in regard to its lack of permanence on a particular place? Is the state permanently stationary and the nomadic continually mobile? Is this what gives the state its sovereign territory and denies the same to the nomad? "The nomad has a territory," Deleuze and Guattari contend.[64] Even de Tocqueville demonstrates some recognition of this claim, for he not only chastises the "Indians" for their "wandering propensities," but for their "instinctive love of country [that] attaches them to the soil that gave them birth."[65]

It was not whether or not the "Indians" had a territory, in the sense of being attached to a particular area that mattered to de Tocqueville, but how they attached themselves to this land. "The Indians occupied [the land]," de Tocqueville claims, "without possessing it."[66] What makes nomads nomadic in the eyes of the state is their orientation to territory. They occupy a rhizomatic territory. They do not "settle permanently." They do not "own property." They do not, to reinvoke the terminology of John Locke, "make use" of the territory. They do not striate the land upon which they live.

Nomads are nomadic due not to their movement, but due to their movement not being the *regulated* movement of the state, and due to their lack of movement not being the striated stasis of the state. The state trajectory *"parcels out a closed space to people*, assigning each person a share and regulating the communication between shares, the nomadic trajectory . . . *distributes people (or animals) in an open space*, one that is indefinite and noncommunicating."[67] Consider de Tocqueville's "childlike Indians," or, on a different plane, "city nomads" occupying the smooth space of the city. They are not allowed to "settle" unless they can properly striate (own) space. Vagrancy . . . loitering . . . no sleeping in city parks . . . Again, it is not movement per se or the lack thereof that constitutes a nomad on Deleuze and Guattari's reading, but the type of movement.

The control of movement is inextricably linked to processes of striation. Cities, oceans, the entire globe have been placed on grids designed to help manage movement. Now, no matter where one is on the entire earth, his/her location can be known. Thus "movement [becomes] the characteristic of a 'moved body' going from one point to another in a striated space."[68] The insistence that movement be a *characteristic* of a body, rather than an *essential part* of it, has significant repercussions for the politics of ecology. The difference with respect to movement is another way of distinguishing the arboreal from the rhizomatic, or the state from the nomad.

For the tree model, the legal model, movement remains confined to specific domains. Bodies occupy specific areas and move in relation to them. Their movements can not be *essential parts* of them; otherwise they could not be said to occupy, or come from, specific points. Thus nomads who move with the seasons, or a particular food source, are without a home, according to the state. They are never settled, according to de Tocqueville. But they are only never settled if movement is merely a characteristic of a body and not an essential part of it. The state's inability to come to terms with this type of movement also suggests an inability to come to terms with ecopolitical problems that very often follow similar trajectories.

Ranging from the effects of massive deforestation, to air or waterborne toxic, radioactive, or other waste products, to the capital that finances ecologically destructive projects, to the activists who struggle against the rapid decline of our environment, to indigenous peoples across the globe, the concept of territory as thought and practiced by the state is problematized by the movements of these actors. The movements of rhizomatic actors deterritorialize by constituting and extending territory in a nonstate fashion. The inability of states to "see" this movement as a part of these bodies rather than a characteristic of essentially stationary bodies forces sovereign state theorists to confine the politics of ecology to the realm of the sovereign territorial state, whether that be through international relations or domestic policy.

The image of the nomad represents more for Deleuze and Guattari than just a different approach to movement. It represents a different form of political community as well. In the nomadic community, Deleuze and Guattari find not a primitive form of political community in the sense that it precedes the state form, or even that it is an incomplete social form awaiting its necessary completion through a social contract, or other such establishment of sovereign government. Rather, they find a political community that anticipates and wards off the state form. Its orientations to movement and to space are just two elements of this.

It is not that these communities lack something that would make them states (government, according to Bull), Deleuze and Guattari maintain; they are constructed so as to prevent the state form from taking hold. Just as "special institutions are . . . necessary to enable a chief to become a man of State . . . diffuse collective mechanisms are just as necessary to prevent a chief from becoming one."[69] The nomadic community is a "not-state" community because its mechanisms are constructed to ward off the apparatus of capture employed by the state form. It could be said that it employs a different model of government, one not based on a model of sovereignty.

Deloria and Lytle provide an excellent description of what Deleuze and Guattari may have in mind by a nomadic form of government in describing the difficulty that Europeans had in coming to terms with the politics of numerous American Indian tribes:

> To the Europeans, Indians appeared as the lowest form of man. No formal institutions were apparent [see Bull's discussion of states presented earlier]. Leaders seemed to come and go almost whimsically. One might be faced with a different person for no apparent reason . . . In tracing the source of political authority, whites were really baffled. No one seemed to be in charge of anything.[70]

The political structure of many American Indian tribes was such that a government in the sense that modern Europeans understood the concept could not exist. The "chief" of a tribe did not hold the reigns of sovereign authority that characterized European kings or prime ministers. Political power for numerous "Indian" tribes was not understood in relation to sovereign authority, as it had been in Europe for centuries.

"There was never a draft in Indian society," Deloria and Lytle add. ". . . if [a] warrior had a good reputation and adventure promised others a chance to distinguish themselves, and if they had confidence in the warrior, then a lot of men . . . would clamor to be a part of the expedition [war party]."[71] Indeed, from Deleuze and Guattari's perspective, this lack of unity between the government of a society and its warmaking apparatus, this different orientation to sovereign authority, was/is a primary weapon in the battle against the state form. The "war machine," as Deleuze and Guattari put it, exists contrary to the state. It resists the striations of the state form, the rigidifications of action and territory. "War maintains the dispersal and segmentarity of groups [so crucial to the maintenance of the state] . . . Just as Hobbes saw clearly that *the State was against war, so war is against the State.*[72] The war machine attacks the boundaries constructed by the state; it breaks them down, returning the striated space of the state to the smooth space of the nomad. In this sense the violence of the war machine differs from the "legitimate" violence that orders the space of sovereignty. The war machine, on Deleuze and Guattari's reading, is essentially antistate.

The method of war making by American Indian tribes, as described by Deloria and Lytle, helps to demonstrate how the state form is repelled by the war machine. As long as such a war machine exists, the state can not exist. States can exist without a war-making apparatus, but in order for sovereign authority to exist in congruence with a war-making capacity, the war machine must become subordinated to the state.

WORLDWIDE MACHINES

In describing the outside of states, Deleuze and Guattari point to not only the local, the minority, the margin, but huge worldwide machines that exceed the boundaries of the territorial state and its corollary interstate system. "Today," Deleuze and Guattari argue, "we can depict an enormous, so-called stateless, monetary mass that circulates through foreign exchanges and across borders . . . constituting a de facto supranational power untouched by governmental decisions."[73] In many ways, this "monetary mass" circulates along nomadic or rhizomatic trajectories. Global capital is not contained by the space of sovereign territorial states.

"It could be said," Deleuze and Guattari claim, "that capitalism develops an economic order that could do without the State. And, in fact, capitalism is not short on war cries against the State, not only in the name of the market, but by virtue of its superior deterritorialization."[74] State sovereignty often gets in the way of capitalism's designs, creating barriers to trade, "nationalizing" industries, enacting environmental regulations, imposing labor safety laws, and so on. Where the state seeks to reterritorialize around a specific point, capitalism deterritorializes across financial networks, trade routes, the internationalization of product manufacturing—producing one part in the United States, another part in Mexico, another in South Korea, another in . . .

The war cries of capitalism against the state are not the whole story, however. The war machine of global capital does not do away with the state, even though it may take charge of the state's aim. "The States, in capitalism, are not canceled out but change form and take on a new meaning: models of realization for a worldwide axiomatic that exceeds them. But to exceed is not the same thing as doing without."[75] States provide necessary preconditions for the creation and circulation of capital; indeed, "capitalism proceeds by way of the State-form."[76] It requires, on one level, the order and stability provided by the state.

Less than one month after the approval of a new constitution in Peru, the Lima office of the Southern Pacific Mail Publications took out an ad, in the 23 November 1993 *New York Times,* claiming that the new government of Peru would not attempt to regulate industries like mining, petroleum production, banking, and insurance, although it would provide basic services such as roads, justice, and security, which would obviously be beneficial to these industries. One section of this two-page advertisement dealing with the new gold rush in the Peruvian Andes mountains reads:

> Five hundred years ago the Incas filled a room in Cajamarca with gold as ransom for Atahualpa, their last great emperor. The [Spanish] conquistadors took the gold and the empire but strangled Atahualpa anyway. It was gold and silver from Peru that funded the great 17th and 18th century expansion of Europe.
>
> Gold, once again, is big in Cajamarca. This time, the foreigners are behaving rather better.[77]

Needless to say, "the foreigners" are behaving better because "Atahualpa" has already been sacrificed to global capital. "The state" no longer stands in the way of the desire of "the foreigners" for gold, or copper, or oil. It helps create the conditions for the most efficient manner of extraction. Said David Lowell, "world-renowned geologist," quoted in the ad about his decision to buy up 120,000 acres in Peru for the purpose of mining copper: "I talked to the army, the Embassy and around 20 taxi drivers. The political risk was down and the Peruvians had put in some of the world's most attractive mining laws."[78]

The worldwide machine of global capital often exists in symbiosis with state space. Although the capital that is now flooding Peru is not contained by the sovereign space of Peruvian territory, it is dependent on and benefited by many aspects of this sovereign space. It requires not only roads, justice, and security, but attractive mining laws as well. It requires a sovereign authority that can maintain "the peace," so that "the people of Atahualpa" won't threaten "the foreigners" again.

"The problem," Warren Magnusson and Rob Walker maintain, for those interested in combating this worldwide machine, "is that capital has no centre . . . Capital itself has never had a centre. There is no heart, no microchip, no source of electricity that can simply be removed to destroy capital."[79] It is a network of relations with no sovereign center. Defeating it, whatever that would mean, in one location does not prevent it from moving someplace else. It is not dependent on a single root, but has multiple roots that extend horizontally rather than vertically. This is the space of ecopolitics. The condition of the earth is bound up with relations of capital.

"Ecology and economy are becoming ever more interwoven—locally, regionally, nationally, and globally—into a seamless net of causes and effects," argues the World Commission on Environment and Development (WCED).[80] The process of humans obtaining their material needs has always had an impact on the earth's ecology. "Today's super communications and larger trade and capital movements have greatly enlarged this process," says the WCED.[81] Mining companies can move from the United States to Peru and not cut themselves off from communications or markets. The ease by which capital can be transferred from one location to another does, as the WCED maintains, endow it with "far-reaching ecological implications."[82] Capital transfers cannot be thought of independently of ecological consequences.

Politics, particularly ecopolitics, must be thought and fought with this in mind. Politics is not confined to state space; and states may well have been appropriated by worldwide machines. Ecopolitics is not transnational, but transversal. It is not reducible to intra- or interstate politics, but slices across the border lines of each, extending itself into different political spaces. Hence, ecopolitics cannot be waged solely within the striated space of the state or its corollary interstate system. It must be waged elsewhere.

CONTESTING POLITICS IN SMOOTH SPACE: GREENPEACE

The task of global ecopolitical thought and action is to confront these new obstacles, paces, and adversaries, and to conceive of new possibilities, avenues, and alliances. Over the past twenty years, global environmental organizations such as Greenpeace have occupied smooth spaces while fighting battles against whaling, nuclear weapons testing, toxic and radioactive waste dumping, and so on, all around the world. Although often pitted against states, Greenpeace's activities have not been confined to the striated space of the state, but have taken place in the smooth spaces of global capital and global ecology as well. In fact, nongovernmental groups such as Greenpeace cannot be located within the striated space of the state. In many respects, they exist outside. Greenpeace could be described as an "international ecumenical organization," to use Deleuze and Guattari's terminology, in that it

> does not proceed from an imperial center that imposes itself upon and homogenizes an exterior milieu; neither is it reducible to relations between formations of the same order, between States, for example (the League of Nations, the United Nations). On the contrary, it constitutes an intermediate milieu between the different coexistent orders.[83]

Greenpeace differs from what international relations scholars usually term an international organization in the sense that Greenpeace's politics cannot be reduced to relations between states as with international organizations housed within the United Nations. Greenpeace's politics are transversal, not transnational.

Although Greenpeace does engage states, and hence does operate within the space of the state, it is at the same time outside the state's structure. When Greenpeace acts, it does not carry the flag of a particular state or group of states. It may well have the backing of people from many different states around the world, but its members cannot be reduced to being citizens of particular states. Moreover, Greenpeace, in keeping with Deleuze and Guattari's elaboration of an "international ecumenical organization," "has the capacity to move through diverse social formations simultaneously."[84] It is able to organize campaigns at a variety of levels, from the local to the global, from the corporate to the national. It operates in a space that blurs the distinction between matters of public and private, national and international, significance.

It is not my intent to idealize or universalize the Greenpeace organization. What I wish to do here is simply point to a few of its actions to draw out the spatiality of its politics. My use of the name Greenpeace will often times be problematic, as I will highlight some events that took place before the name was in place, and other events that happened while the group was so fractured as to make it difficult to decipher which segment could be properly called Greenpeace. Moreover, I do not intend to maintain that Greenpeace is alone in this type of political activity.

Perhaps from the start, Greenpeace recognized the importance and political potential of fighting its battles in smooth space. Many of its best-known campaigns have taken place in the smooth space of the sea, a space that extends beyond the particular territories of sovereign states. In 1971, Greenpeace took to the sea in an attempt to prevent nuclear weapons testing by the United States in the North Pacific. Quaker groups from the United States had been attempting to prevent such tests for years, but had been thwarted over and over by the U.S. Navy. The Navy was able to turn them away and keep the incident contained within the striated space of the United States due to the Quakers being U.S. citizens. When Greenpeace set sail for the Aleutian Islands to protest these tests, the situation took on a slightly different character.

The Greenpeace activists were not U.S. citizens; they were from Canada. Although not official representatives of the state of Canada, the presence of Canadian citizens near the testing sight would cause additional problems for the U.S. naval vessels attempting to thwart the intended protests. Unlike its battles with the Quakers, the Navy could not as easily detain the Greenpeace vessel. The sea, in effect, belongs to no state. To detain vessels from other states would go against international attempts to striate the space of the sea. Greenpeace's protest of U.S. nuclear weapons testing presented a different character to global (eco)politics. This protest could not be reduced to a U.S. domestic problem; neither could it be reduced to a problem of international relations. Greenpeace, and its protest, operated "outside" both state and interstate politics—that is, simultaneously outside the jurisdiction of the United States and the structures of the international system.

Greenpeace's protests of everything from U.S. and French nuclear weapons testing to Japanese, Soviet, and Icelandic whaling serve to deterritorialize state space. They break down the boundaries that the state constructs around itself and its activities. At the same time, Greenpeace's protests reterritorialize smooth space. They extend the space of this nongovernmental organization into a variety of areas, not to take control of it, not to unleash the forces of striation upon it, but to open it up, live in it for awhile, draw it out of the state.

As with the U.S. nuclear weapons tests in the North Pacific, Greenpeace attempted to prevent French nuclear weapons tests in the South Pacific. The strategy was simple: maneuver a vessel into the testing area in hopes that the officials in charge of the test would not risk detonation so long as the Greenpeace vessel was within a fatal range. During one encounter in 1973, French military personnel boarded Greenpeace's *Vega,* which was positioned near the testing site in an attempt to prevent the test from taking place, and beat and abducted members of the crew. Believing that all the film of the event taken by Greenpeacers had been destroyed, French authorities issued reports that it was the Greenpeace members who had attempted to attack them when they pulled up alongside the *Vega* in a small raft to request that it leave the area. When the pictures from the hidden film were published showing French agents bludgeoning the crew of the *Vega* while on the *Vega,* it was France that received a black eye.[85]

But that event pales in comparison with the bombing of the *Rainbow Warrior* by French agents on 10 July 1985, which killed one crew member. The *Rainbow Warrior* was docked in Auckland, New Zealand, on location to stage yet another protest of French nuclear weapons testing in the South Pacific. In an attempt to prevent further protest and stifle the global attention being paid to their activities in the South Pacific, the bombing succeeded only in intensifying the matter.

Greenpeace's attempts to bring the testing of nuclear weapons into the global eye forced France to rearticulate its claims of sovereignty. Immediately following recognition of French involvement in the bombing, French president François Mitterrand issued this statement regarding the future of France's nuclear testing: "The French tests will continue as long as they are judged necessary for the defense of the country by French authorities, and by these authorities alone."[86] French nuclear weapons testing, an activity that has potential global repercussions, is said by this head of state to be the decision of French authorities alone. It is an assertion of French sovereignty into the smooth space of global ecological politics, an attempt to resist the smoothness of the space within which France's nuclear weapons testing takes place, an attempt to reterritorialize the space of French activities around the place of the state.

Greenpeace's protests, however, deterritorialize French territory. They bring out transversal effects of France's "sovereign" decision to test nuclear weapons. The consequences of the decision to test nuclear weapons are not contained by the striated space of state sovereignty. They occupy the smooth spaces of sea and wind currents, which carry the radioactive consequences of France's decision into other territories. They operate within a global ecology that cuts across national boundaries. This deterritorialization of French space is simultaneously a reterritorialization of not-state space. Greenpeace's presence in the South Pacific is not as a representative of a sovereign state in an international dispute with France, but as a nongovernmental organization with members from and connections to many sovereign states and other organizations. Greenpeace's activities reterritorialize in the name of a (potentially) global "citizenry" that exists in no one place.

Greenpeace's use of a worldwide multimedia communications network helps to draw out this process of deterritorializing and reterritorializing. The ability to send audio and video transmissions of its encounters around the world allows Greenpeace to confront states in the smooth space beyond these states. Greenpeace is able to draw its encounters with state vessels and state agencies out of the striated space of the particular sovereign state and transmit them across a global network of radio, television, and print. According to Robert Hunter, an early leader of the group, "the development of the planet-wide mass-communications system was the most radical change to have happened since the earth was created."[87]

This global(izing) media, which allows for the transmission of pictures and stories of activities to people all around the world, deterritorializes as it extends and constitutes territory in a rhizomatic manner; it subverts territories constructed around a sovereign center; it also extends a different territory into and across these sovereign state boundaries; it enables Greenpeace to draw worldwide support for its campaigns. The potential power of witnessing "a whaling ship, an explosive harpoon, a fleeing whale and between them a tiny manned inflatable with the word *Greenpeace* emblazoned on its side—says it all. The image is a godsend for television news, and instantly hundreds of millions of people have shared an experience of

Save the Whales."[88] The extension of Greenpeace support into more and more nation-states, across more and more borders, increases its territory as it deterritorializes the state's. An example of "rhizomatic territorialization," Greenpeace activities constitute and extend territory itself. Following a nomadic trajectory, Greenpeace activists distribute themselves in an open space that is indefinite.

This is a space that capital occupies as well, and Greenpeace (as well as numerous other environmental organizations) is well aware of the potential for battling this worldwide war machine in this space. Rather than focusing solely on state capitals in attempting to battle the worldwide machines of capital, Greenpeace often takes its battle directly to global capital's playing field. The use of consumer purchasing power has been a particularly effective tool of Greenpeace.[89] This strategy targets capital itself. By targeting buyers of particular products, Greenpeace has had significant success, from a boycott of Icelandic fish to prevent Icelandic whaling, to a threatened Alaskan tourism boycott in order to institute a wolf hunting ban, to helping bring about the manufacture of non-ozone depleting refrigerators in Germany by obtaining pledges from over seventy thousand people to purchase such refrigerators. Such activities are not always aimed solely at consumers. The Rainforest Action Network, for instance, organized a boycott of Mitsubishi, a manufacturer of cars, televisions, cameras, and beer to protest its involvement in rain-forest logging. AT&T, Volvo, Boeing, IBM, Apple, and others have been sent materials about Mitsubishi in hopes of convincing these companies to cancel contracts with Mitsubishi. "If we get one company to cancel a mega-million dollar contract with Mitsubishi, we've done more than by convincing consumers," noted Michael Marx, director of the Rainforest Action Network campaign against Mitsubishi.[90]

Before I move on, I want to (re)emphasize one point. I am not arguing that the state has dropped out of the analysis. Obviously states still play important roles in global politics—witness the Greenpeace-France example just discussed. What I am arguing is that the argument that global ecopolitics is a problem solely for separate sovereign states and the interstate system is woefully inadequate. It misses the extent and character of our global ecopolitical problematic. Ecopolitics exceeds the limits of the sovereign state. Yet at the same time, ecopolitical action cannot ignore the state. As Deleuze and

Guattari point out with respect to states in capitalism, "States, in capitalism, are not canceled out, but change form and take on a new meaning."[91] States are still present. They still must be reckoned with, always being aware that the space of ecopolitics is smooth space, not state space.

3

Exploring the Space of the (Inter)State (II): Governmentality

*I don't want to say that the state isn't important; what I want to say
is that relations of power, and hence the analysis that must be made
of them, necessarily extend beyond the limits of the State.*

M. FOUCAULT 1980b

The problem of the space of ecopolitics is not reducible to geography, to questions of territorial boundaries, or even to trajectories.
The problem of the space of ecopolitics also involves questions of
government, or governmentality. To put it another way, the problem
of the space of ecopolitics cannot be reduced to questions of where
territorial lines are drawn and whether or not political issues cross
these territorial lines; the problem of ecopolitics also involves how
these lines are drawn and what is done to the spaces inside these
lines to make them sovereign territories.

For both Waltz and Bull, what is needed to achieve this condition
is authority over a people and a territory. Sovereign territory is territory ruled by a recognized authority that can be differentiated from
other recognized authorities in other sovereign territories. But as I
suggested in chapter 2, there is much more involved in this process
than meets Waltz's or Bull's eyes.

Bull's discussion of the four pillars of the state (population, sovereignty, territory, government), while excluding numerous orientations to sovereignty and government as sufficient to having a state,

treats the concepts of territory and population unproblematically in their relation to the other two. From Bull's perspective, all populations and territories are capable of being states, provided they have governments and are able to exercise sovereignty. Similarly, Waltz unproblematically thinks the relationship between population/territory and sovereignty/government. Both would have done well to pay more attention to the discussion of "government" in the work of one of their traditions' principal voices.

The work of Jean-Jacques Rousseau occupies a prominent position in the tradition of international political thought. Although it is of more importance to Waltz because of the dominant representation of Rousseau's works as fitting more specifically within the realist paradigm, Rousseau's discussion of government provides a strong challenge to the political theory of both Waltz and Bull.[1] Contrary to the realist reading of Rousseau, where he is said to simply accept the relationship between sovereignty and territory/population, a reading of Rousseau's work focusing on his discussion of the problem of government draws out a host of questions concerning the connection between sovereignty and territory/population. Where sovereignty is understood by Rousseau to involve law, government is said to involve economy, the political body's relationship to its population and territory. Ultimately, for Rousseau, the two cannot be thought of independently.

When the problem of government is introduced, the space of ecopolitics is further complicated. It is no longer just about effects of political decisions cutting across sovereign boundaries, or actors that exist outside the space of the state, but it involves how a particular political institution known as the sovereign state relates to its territory and population, how it must shape both in order to become a sovereign state. Through the investigation of the problem of government, what was once a structured, bounded, consistent space, the space of the sovereign, territorial state, becomes a problematic space. Through this investigation, another entry into an ecopolitical theory is provided.

A REALIST READING OF ROUSSEAU

In *Man, the State and War: A Theoretical Analysis,* Waltz attempts to isolate "the causes of war."[2] Rousseau's writings turn out to be invaluable for Waltz in this task. Indeed, "Rousseau's explanation of the origin of war among states is, in broad outline, the final one so long as we operate within a nation-state system."[3] Rousseau's explanation is granted the final word in Waltz's text because "Rousseau himself finds the major causes of war neither in men, nor in states, but in the state system itself."[4] Waltz sorts the various arguments throughout the history of Western political thought for why wars occur into three groups, or "images": the first is an appeal to the evil nature of human beings, the second locates the blame within states, and the third finds the cause of war in the structure of the international system. Rousseau, on Waltz's reading, is a leading theorist of the third image. Rousseau is read as a structural theorist of international politics.

Waltz follows the thought of Rousseau through *A Discourse on the Origin of Inequality*, *Of the Social Contract*, *A Discourse on Political Economy*, and several of his unpublished works on war and peace in the international realm. For Waltz, perhaps the defining moment in all of Rousseau's texts lies in his allegory of the stag hunt. Describing the condition of 'man' in the state of nature, prior to the arrival of civil society, Rousseau relates this thought:

> If it was a matter of hunting a deer, everyone well realized that he must remain faithfully at his post; but if a hare happened to pass within the reach of one of them, we cannot doubt that he would have gone off in pursuit of it without scruple and, having caught his own prey, he would have cared very little about having caused his companions to lose theirs.[5]

"The story is simple," says Waltz, "the implications are tremendous."[6] What Waltz reads into this sentence from Rousseau, which is prefaced by a description of 'men' at a certain stage of development prior to the formation of political entities, where they have "some crude idea of mutual commitments" but beyond that "no foresight whatever,"[7] is a structural analysis of the inherent problems of interacting units in an anarchic system. In a system marked by anarchy, Waltz hears Rousseau telling him that the potential for conflict among the actors is always present. The difficulty of the stag hunt, Waltz concludes, lies not in the particular characteristics of the hunters (their lack of foresight, for instance) but in the structure of their relationship. This example, Waltz contends, "contains the basis for [Rousseau's] explanation of conflict in international relations."[8]

What for Rousseau is an explanation of the long evolution of human characteristics prior to the formation of political societies, becomes for Waltz the paradigmatic description of 'man''s political problem. "We" are self-interested, present-oriented beings by nature, Waltz seems to be saying through this reading of Rousseau. The allegory of the stag hunt is timeless for Waltz. With no overarching authority to mete out punishments to violators, contemporary states are forced to gamble on whether their neighbors will ignore their roles in cooperative environmental efforts and pursue their own economic self-interests instead, in much the same way as one of Rousseau's hunters might give up on the stag and chase the hare.[9] If I leave the hunt to catch the hare, I may ruin the chances for the rest to catch the stag, but my self-interest may well be served. The structure of the system determines the choices for the interacting units, to put it in Waltz's terms—for all times and all places.

For Rousseau, as I read him, the nature of 'men' is far from static. A primary element of *A Discourse on the Origin of Inequality* involves a demonstration of how 'man''s nature has been altered throughout history. One aspect of the problem of government, for Rousseau, is how to make individuals into something capable of thriving in legitimate political communities—citizens. Rousseau recognizes the power of government to shape people. The individual who Waltz accepts as natural is a perversion from Rousseau's perspective, something that is the product of illegitimate forms of political community. But back to Waltz's reading.

Waltz maintains that the analogy between 'men' in Rousseau's state of nature and states in the international realm is accurate as far as the *structures* of the two systems go. But, if we recall Waltz from *Theory of International Politics*: "the two essential elements of a systems theory of international politics [are] the structure of the system and its interacting units."[10] The question of units still plagues Waltz in his reading of Rousseau. The *structure* of the system—states in anarchy—has been provided in Rousseau's thought, according to Waltz, through the allegory of the stag hunt, which can easily be applied to the international arena. But, does Rousseau have anything to say regarding the interacting units that inhabit the international system? Can states, in Rousseau's theory, "be described as acting units?"[11] Without the unified state, Waltz would lose one-half of the necessary elements of his structural theory. Without the unified state, it would make little sense to speak of states as acting in the world. International political theory, as we have come to know it, would be inconceivable. Therefore, if Rousseau's political theory is to be useful for Waltz, it must present states as acting units.

"Fortunately," for Waltz, "it is quite easy to make Rousseau's formulation [of the unified state] concrete."[12] Inquiring about the conditions in which "the state [will] achieve the unity [Rousseau] desires for it," Waltz finds that states achieve unity through public spirit, or patriotism.[13] In a state where public spirit is allowed to flourish, Waltz reads Rousseau to be claiming that "conflict is eliminated and unity is achieved."[14] The state is easily and concretely made into a unit. All it takes is a little patriotism. But, we are soon to find out from Waltz, even if this patriotism is lacking, we can still think of the state as a unit.

"In the ultimate case the unity of the state is simply the naked power of the *de facto* sovereign."[15] Waltz has left Rousseau behind. The appeal to Rousseau's "concrete formulation" has disappeared. The naked power of the de facto sovereign is enough to create the unified state. States are units because there is a legitimate (read de facto) sovereign authority within them.

In *Theory of International Politics* Waltz argues that legitimacy requires only a self-proclaimed right of "the state" to control the use of force in the private realm and the power to back up this right. "The threat of violence and the recurrent use of force," Waltz comments, "are said to distinguish international from national affairs."[16] National politics, it seems, being the realm of authority, adminstration, and law, as opposed to international politics, which is the realm of power and struggle, ought to be the more peaceful of the two realms. Yet history is full of examples that demonstrate that this is not always the case. "*We* easily lose sight of the fact," Waltz confesses, "that struggles to achieve and maintain power, to establish order, and to contrive a kind of justice within states, may be bloodier than wars among states."[17] His examples include Hitler's extermination of six million Jews and Stalin's elimination of five million of his "countrymen."[18] The use of force within states, by states, and, presumably, for states creates a problem for Waltz's theory, of which he is well aware.

The presence of genocidal violence perpetrated within the boundaries of the state by the state apparatus means, for Waltz's theory, that the recurrent use of force cannot be used as a criterion for distinguishing between national and international politics. Where is the unified state in the midst of this intrastate violence? Waltz argues that a distinction can be made on the issue of the use of force between internal and external affairs, however, on the basis of the different "modes of organization" for dealing with force in these two realms. "A government, ruling by some standard of legitimacy, arrogates to itself the right to use force—that is, to apply a variety of sanctions to control the use of force by its subjects."[19] Or, in other words, a sovereign state "has a monopoly on the *legitimate* use of force, and legitimate here means that public agents are organized to prevent and to counter the private use of force."[20]

The sovereignty of each state is (re)enforced by legitimate violence within its territory, and this violent order is then contrasted with the violent anarchy that marks the interstate. In fact, it may even be that the legitimate violence of the state is what gives it its structure. Waltz's move of legitimizing *the* "public" use of force to prevent "private" uses of force does little to confront the problems he mentions regarding Hitler and Stalin. How a state maintains its order is not the structural international political theorist's concern. Internal violence is a problem of national political analysis.

It is on the basis of this separation of areas of inquiry that Waltz responds to another aspect of Richard Ashley's critique of his work from "The Poverty of Neorealism.[21] "Ashley's main objection seems to be that I did not write a theory of domestic politics . . . That is so because I essayed an international-political theory and not a domestic one."[22] He goes on to argue that it is necessary to keep these two fields separate if one is to develop "distinct theor[ies] dealing with the politics and policies of states."[23] But Ashley's charge (and mine as well) is that Waltz *has* written a theory of domestic politics in that he has composed an elaborate argument to validate the presence of the legitimate, sovereign, territorial state. He could not begin to talk about "international-political theory" unless he had already "theorized" the state. Waltz uses the word "assumed" rather than "theorized," and this is no small matter. By claiming to "assume" states as unitary objects, Waltz can appear to have sidestepped any theoretical arguments about the state. Thus he can sweep aside the internal violence that may throw the concept of the sovereign territorial unified state into question.

Internal violence, which takes a multiplicity of forms, is occluded from the international relations theorist's sight by the border line surrounding the sovereign territorial state.[24] Underneath this shroud of legitimacy, drawn over the sovereign territorial state, the state becomes a unified object. Its space becomes the structured space of legitimate violence.

In *Man, the State and War,* Waltz locates this legitimacy in the same place: the naked power of a de facto sovereign. Rousseau is not needed here. Rousseau may even prove troublesome for Waltz at this point. Discussing such theories of sovereignty, Rousseau implored: "Let us agree, then, that might does not make right, and that we are obligated to obey only *legitimate* powers."[25] Rousseau's discussion in *Of the Social Contract* explodes the Waltzian concept of the unified state. For Rousseau, a unified state requires much more than a centralized power to hold all those within certain territorial boundaries in check. A de facto sovereign is not necessarily a legitimate sovereign from Rousseau's perspective.

The further Waltz moves from Rousseau on this point, the better for his theory. In order to construct his theory of international politics, Waltz claims that he "must leave aside, or abstract from the characteristics of units."[26] Even concrete appeals to patriotism could complicate the issue. The units Waltz needs for his structural theory of international politics must be devoid of character. He must leave aside "questions about the . . . social and economic institutions . . . the cultural, economic, political, and military interactions of states."[27] To ask these questions would confuse the issue at hand. Theorists of international politics, at least Waltzian ones, are not interested in what the state looks like. All that is needed for their theories is the assumption that states are units, sovereign entities encompassing territories. On this issue, Rousseau is a problematic voice.

ROUSSEAU'S DISCOURSE OF GOVERNMENT

On Waltz's reading of Rousseau, "the will of the state is the general will; there is no problem of disunity and conflict."[28] Although it may be true that in a state in which the general will is actualized there would be no problem of disunity and conflict, the problem is finding such a state. What is necessary in order to actualize the general will? The answer to this question holds the key for the unified state, according to Rousseau.

Rousseau begins his analysis of legitimacy by quickly dismissing certain states and theories of states that might mask themselves as legitimate, or sovereign. While these two terms mean virtually the same thing for Rousseau in *Of the Social Contract,* they differ markedly from Waltz's use of them in both *Theory of International Politics* and *Man, the State and War.* As mentioned earlier, Rousseau does not find naked power to be a legitimate basis for constructing a state. A de facto sovereign, in Waltz's terms, is no sovereign for Rousseau. Sovereignty, on Rousseau's reading, requires a legitimacy that goes far beyond might.

A legitimate state, for Rousseau, is one that embodies the general will. To repeat Waltz's claim: "the will of the state is the general will." Waltz reduces Rousseau's poitical theory to a tautology, if by state we mean only a political entity in which the general will is actualized. Where Waltz assumes a unified state, or claims that a little public spirit is all that is necessary to create the unified state for Rousseau, Rousseau struggles to articulate principles that would bring such a state into existence. Herein lies the problem of the general will. A legitimate state requires, according to Rousseau, *"each of us* [to put] *in common his person and his whole power under the supreme direction of the general will; and in return we receive in a body every member as an indivisible part of the whole."*[29] Waltz passes over the intricacies of Rousseau's thought here with his conclusion that "the will of the state is the general will; there is no problem of disunity and conflict."

The extent to which Rousseau problematizes theories of sovereignty can be seen in his rejection of even representative democracies as embodiments of the general will. Representative government, from Rousseau's perspective, cannot be said to be legitimate, since the citizens would only be free at the moment in which they vote for their representatives; at all other times they are slaves to the will of others. The will can only be general if all are given the opportunity to ratify the law. In order for the will to be general, the people must be sovereign and sovereignty, like the will, cannot be represented.[30] Rousseau does theorize a unified state, but what is necessary for this unity far exceeds anything Waltz, the *inter*national political theorist, is willing to engage.

There is even a prior problem to that of having the people actualize the general will. First of all, the people must become a people.

Rousseau raises a question that political theorists have long taken for granted. "What people is then suited for legislation?"[31] Rousseau brings into question one of Bull's four pillars of the state: population. Where Bull asserts population as a given, Rousseau suggests that populations appropriate for states must be made. The problem of government is very much encompassed by the problem of population making.

"It suffices to read in chronological succession," argues Michel Foucault, "two different texts by Rousseau": *Discourse on Political Economy* and *Of the Social Contract*.[32] In the former, Rousseau sets out the problem of government, or political economy, as a problem of political practice. What had long been considered private, family, or domestic issues, Foucault argues, exploded as a general political problematic in the sixteenth century. States began to be concerned with the welfare and character of the population, not just its adherence to the law. What had previously been considered the realm of the family, providing for the common welfare of the various individuals (economic and moral concerns), had moved into the realm of the political—that is, the realm of the state. What this problematic of government raised into the space of politics was an entire range of problems from "wealth, resources, means of subsistence, the territory with its specific qualities, climate, irrigation, fertility, etc. . . . [to] customs, habits, ways of acting and thinking, etc.,"[33] things Rousseau maintained were "unknown to our political thinkers."[34]

Traditionally, political theory has been about sovereignty, and sovereignty has been understood as that which is "exercised . . . above all on a territory and consequently on the subjects who inhabit it."[35] For political theorists of sovereignty, "territories [can] be fertile or not, the population dense or sparse, the inhabitants rich or poor, active or lazy, . . ."[36] All of these things are considered to be "mere variables" in the overall problematic of politics.[37] What sovereignty is concerned with, it has been argued in its name, is the demarcation of a specific territory and the imposition of a rule of law over it and the people residing within it. The condition of the territory or the character of the population is not of interest. The problem of government, however, focuses on these variables. And government's relationship to sovereignty, from Rousseau's perspective, cannot be abstracted away.

In *The Discourse on Political Economy*, Rousseau sets himself to the task of developing the principles of government. He begins by asking his readers to "distinguish clearly" between government and sovereignty.[38] Sovereignty is about supreme authority, whereas "everything required by the locality, the climate, soil, moral customs, neighborhood, and all the particular relationships of the people . . . an infinity of details, of policy and *economy*, [is] left to the wisdom of the government."[39] Government takes into consideration all kinds of problems that had been thought inconsequential to the maintenance of sovereignty. It is not enough that government enforce laws that conform to the general will; government must tackle this infinity of details as well.

It is in this light that Foucault reads *Of the Social Contract.* "Using concepts like nature, contract and general will," Rousseau attempts "to provide a general principle of government which allows room for both a juridical principle of sovereignty and for the elements through which an art of government can be defined and characterized."[40] The discourse of government does not eliminate the discourse of sovereignty, nor does it exist independent of it. In *Of the Social Contract,* Rousseau attempts to demonstrate the extent to which they are intertwined. Far from being eliminated by the emergence of government, Foucault contends, "the problem of sovereignty is made more acute than ever."[41]

Rousseau points to a range of problems that complicate the project of state sovereignty. To construct a sovereign state, Rousseau argues, one must first find a godlike legislator to establish the proper political institutions for the society, a properly conditioned people free of enslaving prejudices and customs, a proper territory in terms of size, defensibility, and productivity. After winding through all of the necessary preconditions for sovereignty, Rousseau concludes: "There is still in Europe one country capable of legislation [re: sovereignty]; it is the Isle of Corsica."[42] But given the reality that not all countries are Corsica, Rousseau goes on to provide "guidelines" for approximating the Corsican ideal. These guidelines are not limited to the drafting of an appropriate constitution, but point to the need to "create" both proper citizens and territories.

"It is not enough to say to the citizens, be good; it is necessary to teach them to be so."[43] Rousseau speaks of creating citizens: "To *form* citizens is not the affair of a day, and to have them as men, one must instruct them as children."[44] The enforcement of law and the imposition of punishment are not sufficient toward this end. Government is involved with morals, opinions, customs. If the efforts of sovereignty are to be successful, the character of the population must be taken into consideration. Human beings must be made fit for sovereignty. This is not a simple task, as is evident by Rousseau's claim that Corsica is the only place in all of Europe in the late eighteenth century capable of legitimate government. Presumably, the rest of European peoples had become too corrupt. "He who dares to undertake the instituting of a people," Rousseau warns, "ought to feel himself capable, as it were, of changing human nature."[45] The process of forming citizens, then, can be seen to involve much more than passing a few good laws. But population is not all that must be transformed; territory as well is not a given for Rousseau's theory of sovereignty.

"It is not enough to have citizens and to protect them; it is also necessary to think of their subsistence; and to provide for public needs is an . . . essential duty of government."[46] Rousseau's discourse of government also raises questions concerning a state's interaction with its territory. It is not just about controlling a territory, being sovereign over it, but about relating to it. Rousseau raises the question of territoriality. He is concerned not just with the physical characteristics of the land the sovereign rules, but with the employment of it.

Chapter 8 of book 3 of *Of the Social Contract* deals with questions of climate, production, and surplus and their relationship to forms of government. "When one then asks what is absolutely the best Government," chapter 9 begins, "one poses a question as insoluble as indeterminate . . . it has as many correct solutions as there are possible combinations in the absolute and relative positions of peoples."[47] Government must take into account the subsistence of the citizenry. In doing so, it must take into consideration the climate, the resources, the waterways, the soil, the animals and plants of the territory. Since not all territories contain the same elements, Rousseau finds it impossible to give one answer to the question of which is the best government.

All governments, however, must be concerned with the manner in which the state utilizes its territory. In short, they must seek to transform the territory, if at all possible, for the benefit of the state. Or, as Rousseau puts it, "the end of the political association . . . is the preservation and prosperity of its members."[48] This cannot be accomplished solely through the operations of law/sovereignty. Government involves concerns of a slightly different nature, concerns that have profound ecological repercussions.

What I want to draw out of this discussion is both the way in which the problem of government complicates the project of sovereignty and the way in which the problem of government (and hence the problem of sovereignty) is an ecological problem. By raising the question of government, Rousseau forces thought about politics to come to terms with ecological issues, issues political theorists all too often ignore.

In pointing to the territoriality of the problem of government, I am saying nothing new to the practitioners of politics who have been seeking to transform territories and populations for centuries. But as I argued with respect to the 1864 work of George Perkins Marsh in chapter 1, consciousness of the ecological consequences of government is a rather recent occurrence. Principally only within the last thirty years have states begun to take seriously the ecological repercussions of their governmental activities. Although this contemporary setting will be the subject of the final sections of this chapter, I want to flesh out further this discourse of government and highlight its ecological repercussions through a reading of an eighteenth-century directorate from what was then the governor of "Portuguese America."

THE *DIRETORIO DOS INDIOS*

In 1757, the governor of Portuguese America issued the *Diretorio dos Indios*. On its face, this directorate granted the native Americans residing within Portuguese-American colonial territory full legal rights and called for the transferral of "the temporal authority of the missionaries to secular administrators."[49] The stated purpose of this document was to civilize, or make "citizens" out of, the natives. The process of making citizens was not an end in itself. The American natives were to be transformed into citizens so that they would become "useful to themselves" and "to the State."[50] Their "use" to the state would be exhibited primarily through their taking part in transforming the jungle into more economically useful land forms, thereby contributing to the general wealth of the state. The process of creating citizens did not involve simply granting rights to preexisting "subjects"; rather, it required a creation of "subjects" capable of bearing rights within the state structure. A reading of the directorate's passages reveals the political technologies applied to natives in order to make them into citizens.

To begin with, the "Indian" was to be transformed into a European citizen through an attack on "Indian" culture. Traditional dress was to be replaced with European-style clothes—especially among women. "Indian" social groups were to be restructured into individual family units in part through the construction of European-style cottages with individual rooms. A European-style educational system was begun, where boys and girls were separated and girls would be taught the basic domestic arts of sewing and weaving. Town jails were constructed. Two yearly lists were to be created: the first would contain the names of those who had diligently worked the land, and the second would contain the names of those who preferred to live in idleness[51]—which, when read in opposition to "diligently worked the land," might mean virtually anything else. Although the role of clothes, family structure, educational methods, and penal practices is crucial to understanding the creation of citizens, and helps to point out the extent to which populations are made and not given, the practices I will focus on here involve the territory more directly—the practices of working the land.

The director to be placed on "Indian lands" was "to teach [the "Indians"], not so much how to govern themselves in a civilised way, but rather how to trade and cultivate their lands . . . [which would] . . . make these hitherto wretched people into Christians, rich and civilised."[52] The civilizing process would be carried out primarily through the processes of transforming "Indian lands" into Portuguese farms. The "wretched people" who live in the jungle would become "Christians, rich and civilised," the directorate claims, through the processes of trading and cultivating "their lands." In order to be made into citizens, these "Indians" would have to be convinced of the "dignity of manual labor."[53] Through their labors they would be transformed. Through their labors they would become citizens. By working to transform the land from jungle to farmland or harvestable forest, this "wretched people" would become a "civilized people."

Economic activity, it seems, according to this directorate, is primarily what makes someone a citizen—not to mention a Christian.

But a specific type of economic activity is what is called for in this directorate, one that carries with it a particular orientation to the land. For example, the "casual" method of "harvesting the forest" that was used by the "Indians" was to be replaced by a more "organized" form "so that only the most lucrative commodities [would be] gathered."[54] The shift in farming practices and methods of harvesting the forest were done in order to establish "an opulent and completely happy state."[55] Government.

The issue of the opulence and happiness of the state further highlights the particularity of the practices at use here. The "Indians" had used the forest for centuries before the European invasion. But their methods of "harvesting the forest" were found to be lacking by European standards. In fact, the opposition of diligently working the land and living in idleness suggests that for the authors of the Portuguese American directorate, "Indian" methods of using the land did not count as work, but fell into the category of idleness.

Locke's description of "waste land" is recalled by the rationality of this directorate: "land that is left wholly to nature that hath no improvement of pasturage, tillage, or planting, is called, as indeed it is, waste."[56] On Locke's reading, useful land is only land that has been subjected to modern, European methods of use. All other methods of procuring its fruits and resources are captured in the phrase "left wholly to nature." Thus, "Indian" agricultures are invisible to the Lockean mind, and the work that goes into them is hence also invisible. But to become citizens, the "Indians" had to abandon their work and their agriculture and be convinced of the dignity of "manual labor."

Even the church was found lacking in its attempts to put both "Indian" and land to work. The directorate ordered the temporal authority of the "Indians" to be transferred from missionaries to secular administrators in order to maximize the utility of both the "Indians" and the land. The missionary practice of allowing the "Indians" to remain within missions, working solely for the mission/church, had, "according to the governor, ruined 'the most fertile and fruitful [region] in the world.'"[57] If the transformation of Portuguese America into a state was to be successful, the "Indian" would have to be transformed from a communal, undisciplined hunter-gatherer into a single-family, disciplined, productive unit of the state.

Unlike areas to the north, where vast numbers of Europeans

stood ready to transform the land, thus making the North American "Indian" dispensable,[58] Portuguese America was forced to rely on the "Indian" to carry out this necessary state-forming transformation. Portuguese America simply did not have the numbers of Europeans (already properly disciplined) to carry out this transformation. Thus it did not have the "simple" option of *removing* the Indian from the area so that it could be filled with already "civilized" men and women who would work to make the land useful to the state. In order for Portuguese America to become a part of the state of Portugal, the Portuguese-American "Indian" would have to be "civilized."

The ecological consequences of this civilizing process ought to be obvious. The process was bound up with the transformation of land that had been "left wholly to nature" into land that was put to use. Whether clearing the jungle in order to farm more land, or to create an industry of forestry, the land was to be made by the newly created citizens into a productive resource for the general economy of the state. Where the immediate area had once been put upon by humans to provide for the needs of the local population, the forest would now be put upon to contribute to the general opulence and happiness of the whole of Portugal.

· The needs of the state required not only "the growth of agriculture," but the "civilising of the Indians for the common good," argued the governor of Portuguese America.[59] The two processes are interdependent. Agriculture helps make citizens, and citizens are necessary to conduct agriculture. The destruction of both a jungle and a culture was bound up with this practice of government.

Our attempts today to establish "opulent and completely happy" states still require not only the cultural, and often physical, extinction of many peoples; they continue to require the extinction of thousands upon thousands of nonhuman life forms as well. The transformations that have been carried out around the globe in the name of happy states have carried with them massive environmental degradation. If we are to confront this crisis of ecology, we must come to understand the relationships between the disciplinary mechanisms that operate within and through our political communities and the potential destruction of our life-support system. Our ecological crises cannot be comprehended solely through a juridical sovereign framework, for they do not originate there. The framework of sovereignty within which ecopolitics operates is one that has been complicated by the problem of government. The problem of providing for the welfare of the population has brought the state directly into contact with its territory; and the state's attempts at creating territories capable of providing for its population within the particular standards of this governmentality have greatly contributed to the ecocrises we now face.

As I discussed in chapter 2, this problem is not confined to the territorial boundaries of the state. Not only do the physical effects of these interactions with the earth spill across national frontiers, but the mechanisms of power that are a part of these interactions exceed the limits of the state apparatus as well. The practice of government brought with it, according to Foucault, relations of power that "go beyond the state and its apparatus."[60]

DISCIPLINARY POWER

The practice of government, according to Foucault requires the state to "employ tactics rather than laws, and even use laws themselves as tactics—to arrange things in such a way that, through a certain number of means, such and such ends may be achieved."[61] But how is this "arranging" accomplished? How is the state to see to the production of citizens, the production of wealth, the proper use of resources, the maintenance of a means of subsistence, and so on? In short, what does the practice of government entail? According to Foucault, government, or governmentality, relies on far-reaching and specific disciplinary apparatuses; it involves "the production of an important phenomenon, the emergence, or rather the invention, of a new mechanism of power possessed of highly specific procedural techniques, completely novel instruments, quite different apparatuses."[62] This new mechanism of power is what Foucault has termed disciplinary power.

According to Foucault, power as a concept has long belonged to theorists of law and sovereignty. From this sovereign perspective, power is said to emanate from a central, unified body. Power, on the state level, is about monopolizing the mechanisms of violence in the society. Foucault argues, however, that a new mechanism of power emerged in the seventeenth century, bound up with the problem of government, and quite different from this sovereign understanding of power. Despite the emergence of disciplinary power some two centuries ago, the sovereign conception of power still governs much thought concerning politics today.

For such political theorists, power is something that radiates out from a sovereign center, whether legitimately or not. The problem of power, from this perspective, is very much Hobbes's problem in *Leviathan*, according to Foucault, "the distillation of a single will— or rather, the constitution of a unitary body animated by the spirit of sovereignty."[63] In political thought and analysis, Foucault counters, "we still have not cut off the head of the king. Hence the importance that the theory of power gives to the problem of right and violence, law and illegality, freedom and will, and especially the state and sovereignty (even if the latter is questioned insofar as it is personified in a collective being and no longer a sovereign individual)."[64] There are new mechanisms of power. Political thought and analysis must come to terms with them. These mechanisms of power do not emanate out of a sovereign body; they work through polymorphous disciplinary mechanisms.

"To conceive of power on the basis of these problems [right and violence, law and illegality, freedom and will, the state and sovereignty]," Foucault argues, "is to conceive of it in terms of a historical form that is characteristic of . . . the juridical monarchy . . . For while many of its forms have persisted to the present, it has gradually been penetrated by quite new mechanisms of power.[65] For Foucault, the problem of power is not one that remains confined to a problem of sovereignty as it has been thought by theorists such as Waltz. Relations of power are "employed on all levels and in forms that go beyond the state and its apparatus."[66] The relations of power, the modes of domination we now face are far too mobile, complex, relational to be contained within this sovereign state model of politics. The citizen is not made solely from discourses that emanate from state capitals. Discourses of medicine, psychology, economics, education, agriculture, religion, and so on all contribute to the construction of the proper citizen.

Foucault offers a few methodological precautions for analyzing the power relations at work here: "Refrain from posing the labyrinthine and unanswerable question: 'Who then has power and what has he in mind?'" and instead, "try to discover how it is that subjects are gradually, progressively, really and materially constituted through a multiplicity of organisms, forces, energies, materials, desires, thoughts etc."[67] Foucault sees power as a productive force that permeates and exceeds the state and operates to shape subjects. Try to discover, he suggests, how it is that subjects are constituted in our societies, not just analytically, but really and materially. In short, we should try to discover how it is that disciplinary power operates within society.

In his numerous works, Foucault explores the operations of disciplinary power with respect to how "the individual" is continually constituted and reconstituted. Through studies of madness, criminality, sexuality, and so on, Foucault argues that the individual is a point of intersection on a web of power; the individual is both, and at the same time, constituted by power and its vehicle. As Foucault puts it, to investigate the construct of the individual is to investigate "the individual as an effect of power."[68] What Foucault is suggesting is that the individual is not an originary presence. Instead, the individual is a product of, is constituted by, relations of power. The individual is not a presence that is always already present to receive the instructions of the sovereign. The sane, law-abiding, self-interested, heterosexual, religious, normal individual must be made. And this individual is not made solely by the operations of the state. The power at work on individuals, and populations, goes beyond the state and its apparatus.

Foucault's notion of disciplinary power is extremely helpful, I will argue in the next section, for coming to terms with the merger of Rousseau's government and Deleuze and Guattari's worldwide machines. Government—or governmentality, as Foucault terms it— is not bound by the space of the state. It exists in the space of world-wide machines. It has become global(izing).

GLOBAL(IZING) GOVERNMENTALITY

The governmentality that drove early deforestation in Portuguese America is still at work today. But today, when the Brazilian government sets itself to the task of developing the Amazon rain forest, the process requires the assistance of multinational corporations such as Goodyear, Volkswagen, Nestle, and Borden, and multilateral development banks and international organizations such as the World Bank, the World Resources Institute, the United Nations Food and Agriculture Organization.[69] Now, over two hundred years after the issuing of the Portuguese American directorate, both the effects and the processes of transforming humans and territories are global. The

World Commission on Environment and Development (WCED), issuing its "directorate," asked in 1987: "How can . . . development serve next century's world of twice as many people relying on the same environment?"[70] Where the authors of the Portuguese American directorate were concerned with satisfying human needs and aspirations at the state level, the WCED is concerned with satisfying human needs and aspirations (the major objective of development)[71] at the global level. The mechanisms involved in seeing to the needs and aspirations of the populace have become global. They are no longer confined to the territorial limits of the sovereign state (if they ever were). They constitute, in Deleuze and Guattari's term, a worldwide machine; the mechanisms of economic development encircle the globe, exceeding the state apparatus on many levels.

The problem of government (or development) today differs in another way from that of the eighteenth century. The condition of the earth's environment has taken on a new importance. Where the problem of government in the eighteenth century did involve the territory, the future condition of this territory was not a concern, in the sense of a worry that the very processes used for transforming the territory into a useful object might destroy its utility down the road. As I argued in chapter 1, an ecological perspective that sees human interactions with the earth as potentially devastating to the future condition of the earth did not emerge until the nineteenth century.[72] This ecological concern has become a part of the discourse of government, or development, in the twentieth century. The process of meeting the needs and aspirations of the population in the future depends on there being an environment capable of providing for the needs.

There is, however, an extensive litany of environmentally destructive development projects. Projects to "develop" areas of the Amazon rain forest, for instance—through mining, dam building, and the clearing of land for agriculture and city building—are contributing to problems of polluted waterways, flooding, erosion, deforestation, global warming, and so on.[73] Across the globe, the ever-increasing search for and consumption of energy resources continues to pollute air, water, and land at alarming rates; it continues to cause health problems for populations exposed to the toxins and contaminants given off as wastes through the production and use of these resources. Similarly, the use of fertilizers, pesticides, and herbicides in agricultural production has contributed to the contamination of water, declining health in areas reached by these air- and waterborne contaminants, and the rapid loss of wildlife around the globe. Accompanying this environmental destruction is the cultural and physical destruction of numerous non-European peoples who are either removed from areas deemed necessary to exploit in order to satisfy human needs and aspirations, or transformed into appropriate laborers for this process of development. The practice of government(ality) witnessed in the Portuguese American directorate continues on a global(izing) level today through the practice of development.

It is not enough to point a finger of blame at this monolithic thing known as development. We must attempt to get inside it, come to an understanding of how its mechanisms work. As I have suggested, the apparatus of development is a global(izing) apparatus. It is not something that can be isolated to a particular state. It is perhaps best understood as a worldwide machine. This developmental apparatus is not confined by territorial boundaries. Just as it exceeds state sovereignty in geographical terms, it exceeds the juridical conception of sovereignty with respect to its mechanisms of power. Development needs to be understood from the perspective of a global(izing) governmentality, for it is from this position that we will have the best chance of resisting it, of countering its effects.

The basic rationality for transforming the land and disciplining the population encountered in the Portuguese American directorate of the eighteenth century is a part of many development projects today. But again, these contemporary projects do not have just the welfare of a particular state in mind. Their governmentality is global in scope. The Bastar Tropical Pine Project, for instance, proposed for the Bastar Hills region of India in 1975 by the World Bank, sought to transform the forests of the region into "pine plantations" because of the tropical pine's importance to the global paper and pulp industry. The different trees of the forests (not to mention the myriad of other life forms within the forest) are held up to a standard of utility constructed by the world war machine of global capital. If, as one international forestry consultant puts it, the large biomass typical of tropical forests is "from the standpoint of industrial material supply . . . relatively unimportant (and) the important question is how much of this biomass represents trees and parts of trees of preferred species that can be profitably marketed," then it is easy to see why certain trees become "clearly weeds," and others are farmed.[74]

From an ecological standpoint, it is not as easy to see why diverse forests are being turned into monoculture tree farms. "Life," argues ecologist Daniel Botkin, "is sustained only by a group of organisms of many species—and their environment."[75] Life requires diversity. The move to wipe out this diversity through, among other things, the creation of monoculture tree farms is difficult to defend on an ecological level. By turning diverse forests into monoculture tree farms, in the name of industrial material supply or profit marketability, we are destroying the life-support systems of hundreds of thousands of species that depend on the diversity of the earth's forests. But again, as the forest industry consultant reminds us, from his perspective, from the perspective of the world war machine, these species are "clearly weeds." They get in the way of creating happy and opulent states.

It is difficult to see why some species become clearly weeds from a human perspective as well. Indian environmental activist Vandana Shiva responded to the Bastar Pines project by saying: "It was a project aimed at changing the character of the forests in such a manner that they [would] exclusively serve commercial interests, and not the indigenous peoples."[76] As the Portuguese American directorate sought to transform the forests so that they would serve the state, twentieth-century developmentality seeks to transform forests to serve not only the state, but global capital. Which trees local populations deem useful is not taken into consideration. Their knowledge of both the forest and human interaction with it is not consulted. The utility of the forests in which they live exceeds their use of it. This standard of utility, which hails tropical pines as profitable, simultaneously constructs the other trees of the forests, as well as, perhaps, the people living in these forests, as weeds. This categorization of things as profitable or unprofitable is one aspect of the mechanisms of power of global capital.

The *Tropical Forests Action Plan* (*TFAP*), put out by the United Nations Food and Agriculture Organization, along with the World Resources Institute and the World Bank, and endorsed by the WCED, stresses the importance of *informing* local populations of the benefits they *will* receive from such projects as the Bastar Pines project and of getting them involved in the *implementation* of these projects.[77] Nothing is said of getting contributions from local populations at the planning stages of these projects. Moreover, little is said of the relocation of thousands that will be necessary in order to establish these tree farms.

Perhaps that should not be expected, for, according to this report, "it is the rural poor themselves who are the primary agents of destruction as they clear forests for agricultural land, fuelwood and other necessities."[78] What is needed in order to save the trees of the world, *TFAP* contends, is for forestry to make "the contribution to the well-being of nations in the developing world to the extent [that it is] possible."[79] Otherwise, the rural poor will soon cut them all down. In other words, more "natural" forests must be transformed into "useful" forests not only to satisfy human needs and aspirations, but to ensure the preservation of the forests as well. *TFAP* stands as an example of what the WCED calls sustainable development: development with an environmental spin. According to sustainable development doctrine, development is only possible if the environment is healthy. Forests must be preserved if development is to continue to meet the wood products demands of a rapidly increasing global population. But perhaps the use of the word "forest" here is a bit generous. Remember: some trees are clearly weeds.

Who has lost control of the earth's forests is a question that must not be passed over. The peoples who have lived among the forests of the earth for centuries are rapidly being pushed aside as these forests are turned into useful forests, that is, tree farms. Developmentality requires a shift in control from the local to the global. It requires that "local resources increasingly move out of control of local communities and even national governments into the hands of international finance institutions."[80] In order to meet the needs and aspirations of the global population, local resources across the globe must be readily available to be consumed anywhere else on the globe. But "the logic of international financing" that drives contemporary developmentality responds only to investment return, not to ecological return.[81] Hence, tree farms are set up through loans from the World Bank in order to generate high financial rates of return, divesting the local populations of any access to what used to be their homes, and divesting the land of the diversity necessary to support further growth of not only what the forestry industry considers to be weeds, but oftentimes its "cash crop."

The tree farms that are being created around the globe are often planted on what the World Bank and the United Nations Food and

Agriculture Organization claim is "wasteland," following both John Locke and English colonial policy. "Wasteland" often connotes land long utilized by indigenous populations but not, apparently, properly put to use to meet modern developmentality standards. As Shiva points out, the category "wastelands" was/is a "revenue category, not an ecological category."[82] As a category of colonial policy in India from the seventeenth to the twentieth centuries, "wasteland" referred to lands that were not sources of revenue for the Crown of England. By contemporary developmentality standards, they would be lands not contributing revenue to global financial markets. Thus, forests that provide building, food, and fuel resources for local populations, as well as habitats for animals that are hunted for food and clothing by these populations, can be considered wastelands.

There are over "140 million people who at present live in the earth's forests either as hunter-gatherers, or as swidden agriculturalists or by extracting the produce of the forest on a sustainable basis."[83] These people have not left these lands "wholly to nature." On what grounds can the lands they live in be called wastelands? They have utilized and transformed these lands in ways designed to improve their livelihoods. But "development is viewed as the exclusive domain of production."[84] Land that is not "developed" must be seen as belonging to nature, and "nature is defined as free of humans."[85] "Our" Lockean rationality "overlook[s] in virtually all the accounts of the distribution of the species and the structure of the forests . . . the role of humanity."[86] There is, however,

> a growing body of knowledge on how indigenous and local populations manage their natural resources and sustain them over time. Amplified by the dynamic view of . . . ecological history . . . this knowledge permits an understanding of the forest as the outcome of human as well as biological history.[87]

What this implies is that "natural" forests may be, to a certain extent, human creations.[88] Operating with "Lockean" blinders, however, we must see the "natural" forest as either pristine wilderness or wasteland. Either way, there is a wide range of human interactions with nature that exists outside "our" field of vision.

By recognizing these varying methods of land use "we" can see the particularity of the governmentality at work in transforming "wasteland" into "useful" land, and "natives" into useful citizens;

"we" can see that other methods of land use are available, methods that are not as environmentally destructive as "our" current ones. Moreover, perhaps this will allow "us" to see that other forms of humanity are available as well, forms that are not as environmentally destructive as many of the current forms. In pointing this out, I am not suggesting that "we" all become "natives," whatever that would mean, or that contemporary European economies be replaced with "native" economies. (Why is it that a critique of Western developmental strategies requires such a defense?)

We cannot avoid the worldwide machine of development. We must engage it. And, we must engage it on multiple levels. In the final section, I will turn to an attempt to counter development, not flee from it. What is involved in countering development? On one level it requires a rethinking of the governmentality that drove the Portuguese American directorate and continues to drive twentieth-century development projects. It requires a response to a governmentality that seeks to transform people and territories into useful products for the state. On another level, it requires a response to a developmentality that seeks similar transformations in the name of global capital. The two are not independent of each other. To respond to one may necessarily force a response to the other. On yet another level, countering development requires coming to terms with a politics that is not confined to either the striated space of the state or the smooth space of the world war machine. Politics takes place in both spaces at once. They exist in a mix. In order to counter development, one must recognize the ecopolitical space in which it operates.

COUNTERING DEVELOPMENT

Gustava Esteva, in an essay entitled "Regenerating People's Space," sets himself to the task of articulating a counter to development.[89] Esteva's essay, as it appears in the edited collection *Towards a Just World Peace*, begins with an appendix. He tells a brief story of the "Mexican" people, beginning in the sixteenth century. He relates how nineteenth-century (development) projects turned common peasant and "Indian" lands into private property in the name of transforming these people into citizens of "the so-called 'national society.'"[90] He points out how modern governmental practices have made it possible to "program the body" through a process of making "men and women a quantifiable *population*, a useful *resource*, an *object* of professional attention."[91] Esteva's essay speaks to the problem of government.

Reading development through this lens of governmentality, as Esteva does, helps to demonstrate how it constitutes "nature" and "native" as other, and how it seeks to dispense with them, or transform them into something "useful." It helps to draw out what Esteva calls the "autonomous production of truth, assuming that it is not the production of true statements but of statements through which people govern themselves and others."[92] It also opens a space for rethinking possible human relationships with nature, or, as Esteva puts it, "regenerating people's space."

Regenerating people's space, for Esteva, is a problem bound up with the question "Beyond development, what?"[93] In responding to this question, Esteva first offers a condemnation of development: "'Development' stinks. The damage to persons, the corruption of politics, and the degradation of nature which until recently were only implicit in 'development,' can now be seen, touched, and smelled."[94] Regenerating people's space requires a move away from, or beyond the destruction of development. Esteva is not interested in proposing an "alternative" development. "'We' are opposed," Esteva argues, "to any attempt by the 'alternative' establishment to grant the notion of development a new lease on life through new labels: 'alternative,' 'another,' 'humane,'"[95] and, I would add, "sustainable."

Even when the literature produced by establishment organizations across the globe, including the United Nations, has pointed out the severe damage wrought by development, the conclusion has not been to abandon development, or even to seriously rethink it. Rather, it is assumed that there must be a "missing factor or tool, or perverse, corrupt, or inefficient use of something [that] explain[s] the damage done by development to people and their environments."[96] Esteva questions the inability, or refusal, of these analyses of development to question development at its core. This inability or refusal can be seen through a reading of the United Nations World Commission on Environment and Development's 1987 report, *Our Common Future.*

Our Common Future, more popularly known as "The Brundtland Report" for the prime minister of Norway who chaired the commission, is perhaps best known for developing the term "sustainable development." In the words of Jim MacNeil, the principal author of

Our Common Future, sustainable development is "not the type of development that dominates today, but development based on forms and processes that do not undermine the integrity of the environment on which they depend."[97] Recognizing that development cannot exist on a dead planet, the WCED attempts to weave together the fields of economics and ecology, making a case for a new model of development that will take seriously ecological necessities.

In a section titled "The Interlocking Crises" the WCED argues powerfully that environment and development are not separate challenges, that "ecology and economy are becoming ever more interwoven—locally, regionally, nationally, and globally—into a seamless net of causes and effects."[98] Underscoring MacNeil's claim that sustainable development requires a concern for the integrity of the environment, the WCED underscores the connection between development and environment. "Development cannot subsist on a deteriorating environmental resource base."[99] Sustainable development will differ from other forms of development, we are told, because of the recognition that the process of development depends on a healthy environment. But a closer reading, one that focuses on the question of poverty's relationship to environmental degradation and control over resources, will point to Esteva's general criticism of "alternative" forms of development.

Throughout the process of researching and writing this report, its authors claim that one central theme emerged: "many present development trends leave increasing numbers of people poor and vulnerable, while at the same time degrading the environment."[100] Despite this "admission," the report offers no sustained critique of development as such. The damage done to the environment by "many development trends" would seem to be an obvious starting point for the WCED in its analysis. It would be a point from which an analysis of the relations of power present in development strategies and processes could be launched. Yet, in the face of this "one central theme," the text goes on to endorse development as an unquestioned economic and environmental necessity.

The problem as it is articulated by the WCED is not how to avoid the poverty, vulnerability, and environmental degradation wrought by development, but rather, "How can such development serve next century's world of twice as many people relying on the same environment?"[101] The goal quickly becomes to "make *development* sustainable."[102] Very little is offered in the way of an examination of the degradation to the environment that development has wrought.

The question of poverty's connection to environmental destruction is vital for this particular recasting of development. As long as poverty is presented as the cause of environmental destruction, sustainable development can enter the arena unscathed—witness Shiva's critique of *TFAP* mentioned earlier. Since all development supposedly aims to eliminate poverty, sustainable development can now claim, as an additional aim, the elimination of environmental degradation. To see this move it is necessary to reprint two paragraphs that form the lead-in to a section on "Causes and Effects" in *Our Common Future*, two paragraphs that remove the professed "one central theme" from view:

> Environmental stress has often been seen as the result of the growing demand on scarce resources and the pollution generated by the rising living standards of the relatively affluent. But *poverty itself* pollutes the environment, creating environmental stress in a different way. Those who are *poor and hungry* will often destroy their immediate environment in order to survive: *They* cut down forests; *their* livestock will overgraze grasslands; *they* will overuse marginal land; and in growing numbers *they* will crowd into congested cities. The cumulative effect of these changes is so far-reaching as to make *poverty itself a major global scourge.*
>
> On the other hand, where *economic growth* has led to improvements in living standards, it has *sometimes* been achieved in ways that are globally damaging *in the longer term.* Much of the improvement in the past has been based on the use of increasing amounts of raw materials, energy, chemicals, and synthetics and on the creation of pollution that is *not adequately accounted for* in figuring the costs of production processes. *These trends* have had *unforeseen effects* on the environment. Thus today's environmental challenges arise both from the lack of development and from the unintended consequences of some forms of economic growth.[103]

In this lengthy passage we can begin to get a sense of the extent to which the one central theme of development's role in the creation of poverty and environmental degradation has left the analysis of the WCED. Moreover, we get the WCED's answer to the question of poverty's connection to environmental destruction. *Poverty* emerges as a principal destroyer of the environment. Its position as a possible *consequence* of development is not examined. How the poor come to their poverty is not examined. How a degraded environment creates poverty, or what role development might have in this process, is not examined. Poverty itself pollutes, we are told. The poor cut down trees, allow their livestock to overgraze grasslands, crowd into overcongested cities, and so on. Poverty in and of itself is a major global scourge.

On the other hand, economic growth sometimes has unforeseen effects on the environment in the long term. There is no cutting down of trees, no overgrazing of grasslands, no congesting of cities by economic growth. No, the sins of economic growth are trends that have had unforeseen effects on the environment. The use of increasing amounts of raw materials, chemicals, and synthetics, and the creation of pollution that accompanies economic growth, while increasing standards of living, have not always been factored into the costs of production. This is the problem with environmental growth.

The shift in language from the paragraph dealing with poverty to the one dealing with economic growth is revealing. The former contains an active voice. An agent is clearly identified (the poor) and is fixed with responsibility for the environmental degradation that results from its action. In the later paragraph, no agent is clearly identified. Economic growth, unlike poverty, does not reveal a particular actor. While the first sentence of the quote does begin to suggest the role of the relatively affluent in the destruction of the environment, the language used is "has often been seen," and it is followed with the inevitable "but" that affixes responsibility to the poor. Economic growth has generated pollution, the WCED allows, but the problem is not with the pollution itself; rather, it is with the failure to factor this pollution into the processes of production that have led to improved living standards. But poverty itself pollutes.

It is perhaps more than an argument based on semantics to point out that the sign development escapes association with environmental degradation in these paragraphs. The second paragraph is written in such a manner as to keep development unsullied by the current condition of the earth's environment. "Economic growth" is used in the two spots where some blame for the current condition of the environment could be attached to development if development and economic growth are thought of as synonymous. This is necessary if "today's environmental challenges" are to be presented as arising "from the lack of development." By keeping the sign of development pure, and by arguing that poverty is a cause of environmental destruction, development (in this case, sustainable development) can emerge as an environmental necessity. The implications for the world's impoverished are perhaps obvious.

Saddled with the blame for the current condition of the environment, attempts to bring them out of their poverty can now be argued to be in the interest of the health of the earth's ecology. What this entails, again as Shiva comments in her critique of the WCED-endorsed *TFAP*, is a continued intervention into the local population's control over land. If "the poor" are to be blamed for destroying tropical rain forests, let's say, then it makes sense to remove them from these lands, particularly if it is to be done by development; for poverty, which is presented as the destroyer of the earth's forests, pasturelands, and so on, becomes the principal cause of development's inability to sustain itself. If it were not for the poor cutting down trees, allowing their cattle to overgraze grasslands, and so on, development would have the resources necessary to sustain itself; moreover, it would be able to sustain the integrity of the environment. But this claim fails to take into account that one central theme that "many present development trends leave increasing numbers of people poor and vulnerable, while at the same time degrading the environment," and thus, from Esteva's perspective, grants development a new lease on life through the use of a new label.

Widespread acceptance of at least the idea of development, Esteva acknowledges, makes articulating a non- or counterdevelopment discourse of governmentality difficult at best. "Indeed we get incredulous looks when we say that all these forms of 'alternative' development are nothing but a deodorant to mask the stink of development . . . The stark fact is that we are not taken seriously unless we accept some version of 'development.' "[104] "It is only by saying no to development (i.e. to the power it creates and the language that supports it)," Arturo Escobar argues in "Discourse and Power in Development," "that Third World countries can fruitfully attack adverse conditions with new discourses and knowledge, new ways of fulfilling basic needs, of realizing the possibilities of human beings."[105] If "many development trends" are responsible for environmental degradation around the world, perhaps it is development, and not the specific trends, that needs to be attacked.

Esteva says no to development. In saying no he seeks to regenerate a space in which a new discourse can be articulated, where new ways of fulfilling basic needs can be suggested, where new avenues can be pointed to for realizing possibilities for human beings. Within that space he offers a one-word response to the question "Beyond development, what?"—"hospitality."[106]

The choice of hospitality as a response is intended to cut to the core of development. Development, as a recent expression of the Western ethos, according to Esteva, views the world as inhospitable. Hence, the environment must be transformed from wild, teeming, dangerous nature into controlled, ordered, useful land. This is necessary, the development plea goes, if "human" life is to take place. The image of hospitality, however, takes into account the fact that if the world were inhospitable to humans, we would have perished long ago. Hospitality emerges from a very different ethics, one much closer to the Nietzschean ethic I tried to draw out in chapter 1. Operating within an ethic of hospitality as set out by Esteva provides a very different view of the world.

The development view of the world, according to Esteva, is universalist. Values that this discourse takes to be universal (as is seen in the presentation of "human needs and aspirations") the discourse of hospitality sees as culturally relative. Hospitality operates within this realm of "cultural relativism," Esteva argues. It implies "the dissolution of universal values."[107] But before this can send too many chills down the spine of his moralistic readers, Esteva continues: "This does not mean, of course, having no guiding principles to live in community. It means exactly the opposite: to have them fully rooted in the perception and attitudes of daily life, instead of supplanting them with artificial constructs which are hypothetically universal and more or less ahistorical."[108] Esteva attempts to write a politics of daily life. He seeks to challenge the universalist governmentalities that refuse difference and seek to cast it out, just as the trees that are clearly weeds are to be cast out. Hospitality offers a wider range of useful; yet it does not force bodies to become so in order to live.

The ethic that accompanies Esteva's notion of hospitality is similar in important ways to the ecological ethic of care I found in Nietzsche's thought. Both contain notions of taking care not to "divest nature of its rich ambiguity." Both offer reorientations to otherness that allow for a reverence for difference rather than an eradication of it. Both challenge notions of political space as confined to the space of the state. Nietzsche points to the border of the state and Esteva seeks to regenerate people's space.

Hospitality as a response to developmentality, however, is insufficient, Esteva cautions. Hosting the inhospitable can be a deadly game. By hosting the Spaniards, the "people" were colonized, Esteva claims; by hosting Western gods, their gods were destroyed; "by hosting 'development,' their environment and livelihood were seriously damaged."[109] "Because of the damage done by development, it is not enough to be hospitable . . . 'we' need to apply *remedies*."[110] Efforts must be made to counter the effects of development. Efforts must be made to confront the political structures of development. This may mean that ecopolitical actors have to deal with state, interstate, or development agencies. But, Indian environmentalist Anupam Mishram cautions,

> we should not assume that we can look for solutions to our problems within the framework of the current development pattern. It would be folly to think the Brundtland commission can find solutions within the "counter-productive framework" of governments, the United Nations, the World Bank and so on. Because the present structures have given us the disease, is it then logical that they should also provide the cure?[111]

Responding to development requires activities that break from the actors that have brought it to us. So what solutions, what programs are available? Esteva offers no global or universal strategies for such engagement.

He refuses to offer a program for fear that it will become universalized. He comments briefly on certain projects that he has been a part of in Mexico, but he provides no substance to these projects. To do so, in his mind, will only end up duplicating the very rationality he is struggling against. The world of ecopolitics is far too varied, contingent, and ambiguous to think that a project that had some success in a town in Mexico will necessarily be successful elsewhere. The machine of development is worldwide, but it is also amazingly specific. Battles must be waged on a variety of fronts, and in a variety of ways.

Esteva takes seriously the differences that cut within and across environments and cultures. His intent is not to "propose, or impose a 'global design' for the national or international arena."[112] He calls for approaches to the problem of government that emerge from the environmental and cultural "realities" of the particular localities. "A 'Global' perception that cannot be expressed in local terms (in all possible locations) lacks true reality. It is mere speculation."[113] The problem of the global environment cannot be used to abstract away the local effects of global ecopolitics. The problem of global deforestation is not independent of localized struggles over resource use and control from India, to Malaysia, to Brazil, to Canada, to . . .

Moreover, despite the presence of worldwide machines, "the political regime that is characteristic of our time has *politicized daily life*."[114] "We" have all become points of developmentality's application. Ecopolitical struggles over resources "over there" are tied to consumption patterns "here." And, in keeping with the governmentality of the Portuguese American directorate, daily life everywhere can be tied to the disciplinary power of our global(izing) governmentality. But just as global perceptions cannot ignore the local, Esteva cautions against overprivileging the local.

"The 'local' perception that cannot see itself in 'global' terms and dimensions," Esteva adds, "fails to see itself with sufficient depth."[115] Just as the problem of global deforestation cannot be detatched from local struggles over access to forests, local struggles cannot afford to see themselves as separate from global processes. The not-in-my-backyard (NIMBY) strategy comes to mind here. Local attempts to isolate or withdraw one community from the polluting aspects of a particular industry are often ignorant not only of global economies but of global ecologies. "There are no groups, peoples, ethnic communities, cultures or societies," Esteva warns, "living without 'contact' with the 'outside.'"[116] And, as with Delueze and Guattari, the "outside" can be understood to exist across many spaces and throughout many locales.

Esteva seeks to articulate a local politics with a global orientation, a global politics with a local emphasis. His attempt to move beyond development leads (back) into the question of space. A blurring of local and global takes place here. This is perhaps no more Esteva's prescription for the state of contemporary global politics than it is his description of the world in which late twentieth-century developmentality takes place. As with Deleuze and Guattari's analysis of rhizomatic space, Esteva's attempt to regenerate people's space is based on an understanding of a new political problematic of space. Developmentality is not confined to a world striated into separate sovereign territories known as states. Attempts to respond to it cannot afford to confine themselves to these spaces. By confining responses to the worldwide machine of development to the space of the state, whether that be geographically or institutionally, ecopolitical theorists fail to recognize the scope and difference of ecopolitics. The space of ecopolitics, I have been insisting, exceeds the striated lines of the state in multiple directions.

4

Theories of Ecopolitics:
Machines, Organisms, Cyborgs

We are living through a movement from an organic, industrial society to a polymorphous, information society . . . from comfortable hierarchical dominations to scary new networks . . . of domination.

D. HARAWAY 1991c

The subjects are cyborg, nature is coyote, and the geography is elsewhere.

D. HARAWAY 1991a

Has the field of ecopolitical thought come to terms with the critiques of state space set out in chapters 2 and 3? Ecopolitical theorists have long maintained the radicalness of their field of thought, claiming to break from much of what formed the mainstream of Western thought dating as far back as the emergence of monotheism, the creation of rationalistic philosophy, the birth of experimental science, and so on. But where does this field stand in relation to the question of rhizomes, nomads, worldwide machines, governmentality? In short, where does it stand with respect to the space of ecopolitics as I have mapped it?

I will begin this discussion by admitting to the problematic move of treating the widely disparate work relating to ecopolitics under the rubric of "a field of thought." If for no other reason, works ranging from Aldo Leopold's "Land Ethic" to Susan Griffin's *Woman and Nature* can be said to occupy a field of thought because of their concern with humanity's relationship to the nonhuman world. Beyond that, many of these works have little in common. Recognizing the inherent problem of attempting to situate disparate ecopolitical works and thinkers on a field, Andrew Dobson defends his attempt in *Green Political Thought* by claiming that such a move "makes clearer the territory within which [ecopolitical thought] most properly moves."[1]

Stating that it would be difficult to improve upon the effort by Tim O'Riordan in his 1976 work *Environmentalism*,[2] Dobson goes on to re-create the ecopolitical thought matrix drawn by O'Riordan. Utilizing axes of sovereignty and territory, O'Riordan and Dobson divide the field into four camps:

Size	"new global order" WCED Ward and DuBos	"centralized authoritarianism" Ophuls Hardin
of ↑		
Territory	"anarchist solution" Bookchin	"authoritarian commune" Heilbroner Goldsmith?

Centralization of sovereignty →

Under the heading "new global order" they list ecopolitical theorists who call for a loose coordination of attempts to manage the global environmental crisis, perhaps through the United Nations or some similarly quasi-federated international organization.[3] Under a "centralized authoritarianism" banner they place ecopolitical theorists who suggest that people will not willingingly adopt a more ecologically friendly relationship to the earth and hence must be made to do so by state institutions.[4] Ecopolitical theorists who argue that institutional structures must be localized, yet maintain an authoritarianism for the reasons suggested above are claimed to be supporters of an authoritarian commune.[5] Finally, theorists who find authoritarianism as well as large sovereign territories contrary to ecological principles and argue instead for local decentralized communities are placed under the "anarchist solution" heading.[6]

O'Riordan's and Dobson's matrix leaves many issues crucial to numerous ecotheorists off or hidden within the field of discourse. Primary in this regard are conceptions of "man" and nature.[7] By dividing the field of ecopolitical thought up as they do, O'Riordan and Dobson make a claim about the space of (eco)politics. Sovereignty is held to be the primary political question, and territory is presented as an object predisposed to the exercise of sovereignty. The issue of humanity's relationship to nature, we must assume, is either excluded from the realm of the political, or subsumed by the larger questions of sovereignty and territory, or, since their matrix involves organizing ecopolitical thought, felt to be the same across all inhabitants of this space.

What is intriguing about this matrix is the extent to which it locates itself solidly within that long tradition of thought about politics examined in chapters 2 and 3. Sovereignty and territory frame this matrix. Despite Dobson's claims that true "ecologism," or green political thought, involves a rejection of the state as the place of politics, when it comes to orienting the field of ecopolitical thought, he orients it around the twin towers of the modern Western state: sovereignty and territory. Ecopolitics stands in opposition to state politics, from Dobson's perspective. But, by privileging sovereignty and territory as the axes for his matrix of ecopolitical thought, he reifies state space as political space.

A look inside this matrix will help to draw out my argument. Since Dobson opposes true "ecologism" (what he claims occupies the "anarchist solution" square) with state politics, it will perhaps be useful to examine both an example of green political thought, which, in Dobson's eyes, rejects the state model the least, and an example that rejects the state model the most. The works of William Ophuls and Murray Bookchin will be examined in the next two sections.

Ophuls and Bookchin can be read as contrasting thinkers in numerous ways (which I will attempt to demonstrate through my reading of each). But despite these differences, when viewed from a position off the field as O'Riordan and Dobson have constructed it, a number of critical similarities come into view. Among them are similar orientations to the politics of sovereign territory and similar understandings of nature and human beings. Neither author engages the problems of nomadic trajectories and worldwide machines, governmentality, and disciplinary power, or even the death of God and the lingering presence of the shadow of God set out in the previous three chapters. Hence their works fail to address certain aspects of the space of ecopolitics that I have been arguing are crucial. After presenting readings of each, I will turn to an ecopolitical theorist who, I will argue, does engage these problems.

If Ophuls is a mechanistic theorist, and Bookchin is an organic theorist (to use these old political metaphors), then the work of Donna Haraway is properly cyborg.[8] By this I do not mean a compromise, or synthesis of the two, but more directly a de(con)struction of machine and organism as ruling metaphors for social and political thought, something that breaks down the boundaries between the two and creates something new. Where Ophuls and Bookchin are unable to come to terms with the (post-)Nietzschean reconceptualizations of nature, humanity, and politics, examined in the preceding chapters, Haraway's cyborgism not only draws on such reconceptualizations but it articulates a possible ecopolitics that flows from them.

OPHULS AND THE PROBLEM OF SCARCITY

> *In brief, liberal democracy as we know it . . . is doomed by ecological scarcity . . . the whole ideology of modernity growing out of the Enlightenment, especially such central tenets as individualism, may no longer be viable.*[9]

In *Ecology and the Politics of Scarcity*, William Ophuls makes a forceful argument that the ideology of liberalism/modernity is ill equipped to handle the ecological crises we now face. Speaking primarily of the control, use, and availability of natural resources, Ophuls argues that the conditions for thinking about politics have changed drastically. The governing liberal/modern paradigm, of which John Locke and Adam Smith are founding voices, is based on assumptions of ecological abundance, assumptions that are no longer valid.

The "bonanza" of newfound wealth in the Americas, Ophuls maintains, contributed greatly to the modern doctrine of liberalism and its embrace of such principles as democracy, freedom, private property, and individualism.[10] The late-modern world, however, is marked by a scarcity that will "necessarily undercut the material conditions that have created and sustained current ideas, institutions, and practices."[11] In short, Ophuls argues, we are entering a new era marked by ecological scarcity, necessitating new political ideas. Modern Western liberal society, Ophuls maintains, is "deliberately set up to . . . encourage the ruthless, competitive exploitation of the commons."[12] A reading of one of its founding voices will prove beneficial to drawing out Ophuls's position here.

The liberal society constructed by John Locke begins, as Ophuls suggests, with particular assumptions about natural abundance. According to Locke, the world had been "given to Men for the Support and Comfort of their being" by God.[13] Beginning with this assumption, Locke constructs a theory of politics grounded in a defense of property rights. Although Locke did maintain that God gave the world to "mankind in common," he went on to argue that this does not mean that God meant it to go unused, or to remain in common forever.[14]

Claiming that we have a property in our physical persons, Locke takes a short step to claim that what we put our labor to becomes our property as well—an extension of our physical being. To protect against one person who by picking all the apples from a tree establishes a property over them and thus deprives all others the fruit of this tree, Locke adds a proviso that one can take only as much as any one can make use of "to any advantage of life before it spoils."[15] In conditions of scarcity, this proviso would still not ensure all access to enough to sustain themselves, but from Locke's perspective this was an adequate proviso due to the presence of an overabundance of land not being properly put to use (especially in the Americas).

As I have pointed out more than once, Locke considered land that is "left wholly to nature that hath no improvement of pasturage, tillage, or planting" to be waste.[16] The presence of large tracts of wasteland across the earth meant that all could put their labor to the land, claim property for themselves, and deprive no one of a means of subsistence. Scarcity and poor conditions of life existed, according to Locke, where the people did not properly put the bounty of nature to use. This difference in approach to land use was *the* distinguishing characteristic for Locke between Europe and America.

Societies that labored (as Locke understood the term) would prosper; those that did not, would not:

> There cannot be a clearer demonstration of anything, than several Nations of the *Americas* are of this, who are rich in land, and poor in all the Comforts of Life; whom Nature having furnished as liberally as any other people; with the materials of Plenty, i.e. a fruitful Soil, apt to produce in abundance, what might serve for food, rayment, and delight; yet for want of improving it by labour, have not one hundredth part of the Conveniencies we [England] enjoy.[17]

Where land in America went to waste, in Europe it was put to use. As I demonstrated in chapter 3, the concept of "wasteland" would function as a justification of British conquest across the globe. The "natives" the British encountered could not possibly "own" the land they lived on since they were not properly putting it to use. Property is a product of labor; and according to Locke (and many Westerners then and now), most of the earth was not worked by human hands and hence belonged to no one. It was England's for the taking. According to Vandana Shiva, "when the British established their rule in India, it was estimated that between one-third and one-half of the total area of Bengal Province alone was 'waste.' "[18] The land's official status as waste opened the door to British colonizers, or perchance the "civilized" Indian, who would be willing to clear the forests, drain the marshes, and fence in the land for cattle or cultivation. The people who had been living on this "wasteland" were swiftly removed—they had no right to the land; they had not put it to use.

Starting with what he considered to be a natural right to private property, Locke went on in his treatise on civil government to set out an institutional framework for a representative government that will protect not only an individual's right to property, but other individual rights as well. "Our problem" today, Ophuls argues, "is not so much that our [liberal] institutions no longer work the way they should . . . but that the assumptions about the carrying capacity of the commons which supported these institutions are no longer true."[19] Locke's conceptions of use and the natural right to property through labor are not critiqued by Ophuls. Ophuls focuses solely on what he calls Locke's "cornucopian assumptions," and hence the libertarian ecopolitics that emerges from Locke's theory. The issue for Ophuls becomes rectifying property based on labor with conditions of scarcity. We do not need new political institutions to solve this problem, just a recognition in these institutions of the existence of scarcity.

What this entails, according to Ophuls, is "more government—that is, stronger checks on the competitive overexploitation of the ecological commons and therefore on human aggrandizement."[20] Without government taking on this role, Ophuls maintains that we will only perpetuate what Garrett Hardin has described as "the tragedy of the commons."[21] What is tragic about the condition of the commons, according to Hardin, is that each person, pursuing his or her own self-interest, will use up more and more of the resources of the commons for fear that others are doing the same thing. Hardin sees in this tragedy a powerful argument for an "ecology" of private property. It is only when each self-interested individual actually owns a particular plot of land, water, air, and so on that he or she will be concerned about conserving the resources in it.

As with Hardin, Ophuls promotes an ecology of private property, but unlike Hardin, he goes on to argue that private property is insufficient in the face of the ecological scarcity that marks the modern world. Locke's spoilage proviso—not to mention Hardin's eco-self-interest—is insufficient to protect the ecological commons. Self-interested individuals, left to their own devices, Ophuls fears, will not preserve the earth's resources even if they acquire a property in these resources. Although he hopes that we will learn ecological restraint, the bulk of his work suggests a realization that we will not be able to do so without the force of government.

He goes on to argue for "the erection by a majority of a sovereign power that would constrain all men to be reasonable and peaceful."[22] In short, the division of the commons into private property will not be/has not been sufficient to protect the supply of natural resources or the quality of the environment. The tragedy of the commons is not necessarily abated by privatization. A sovereign authority will be necessary to legislate the temperance that may be lacking within individuals.

As I mentioned earlier (see note 4 to this chapter), Ophuls has taken exception to the reading of his work as advocating an ecological Leviathan. Although he does suggest in his 1973 essay titled "Leviathan or Oblivion?" that a Hobbesian politics would be better equipped to handle our current problem of scarcity since scarcity was Hobbes's main concern,[23] he later claims that "we need a form of government that is effective in obliging humankind to live within its ecological means," but this does not mean constructing some type of ecototalitarian regime.[24] Although this is not the place to engage this issue fully, Ophuls's repeated contentions in his 1977 text that his emphasis was on legislating temperance, not creating new bureaucracies, should have been sufficient to stave off interpretations that he is an advocate of an overly administrative approach to the ecopolitical problem. Yet, from the perspective of a Dobson, who sees eco-anarchism as the only true ecopolitical theory, Ophuls's appeals to "more government," although not "more administration,"[25] are enough to place his work in a "centralized authoritarianism" camp.

Ophuls sees the need to have government legislate temperance, or self-restraint of the will, upon individuals if they are unable to do so themselves. In this sense, Ophuls should not be seen as engaging a Rousseauian discourse of government. In recognizing the need for a government effective at "obliging" people to live within the earth's ecological limitations, his analysis is not suggestive of the productive aspect of Rousseau's notion of the role of government, or Foucault's concepts of disciplinary power and governmentality. Ophuls does not see the very self-interested individuals who must now be restrained from overexploiting the earth's resources as products of a particular governmentality. Along with Locke, he accepts as natural much of what is contained within the 'economic man' of liberalism. Thus, if this person should be unable to restrain his/her own over-exploitative behavior, government will have to be strong enough to check it.

In this move, Ophuls attempts to (re)claim the space of state sovereignty in the face of liberal individualism. His argument that the sovereign authority he advocates is democratic notwithstanding, the primacy of the space of the state is reasserted. The state itself does not come under question; only its actions do. Nonetheless, Ophuls's ecopolitical theory does elicit an appreciation of the extent to which ecopolitics exceeds the boundaries of sovereign state territory.

While Ophuls does recognize the insufficiency of confining ecological politics to separate sovereign territorial state borders, the only way he can see to deal with this problem is to extend the reach of sovereign state institutions. Following the "state-of-nature" analogy that has become famous in international theory,[26] Ophuls warns that

> without some kind of international government machinery with authority and coercive power over sovereign states sufficient to oblige them to keep within the bounds of the ecological common interest of all on the planet, the world must suffer the ever greater environmental ills ordained by the global tragedy of the commons.[27]

Presenting the existence of sovereign states across the globe as analogous to sovereign individuals within a state of nature, Ophuls argues that a *global* sovereign would be necessary to solve the global ecological crisis. Where Waltzian (neo)realists argue that the position Ophuls is advocating is nothing more than fanciful idealism, Ophuls might respond that the (neo)realists are the ones being idealistic, if they think "balance of power" and diplomacy will work to solve the global tragedy of the commons. If states-as-actors are self-interested, then Waltz's claim that global ecological problems depend on national solutions is an idealistic dream given the reality of Hardin's tragedy of the commons scenario. The proper global government machinery must be constructed.

In a section titled "Planetary Government or the War of All against All," Ophuls argues that there exists a "strong rationale for a world government with enough coercive power over fractious nation states to achieve what reasonable men would regard as the planetary common interest."[28] While acknowledging the "overwhelming" need for such a global authority, Ophuls adds that the ecological scarcity that makes a world sovereign necessary also makes it more difficult to achieve.[29] Those states that are in a somewhat more enviable position will be less likely to cater to the *planetary* common interest until, perhaps, it is too late. Rather than providing the appropriate motivation to form a world government powerful enough to enforce the necessary ecological laws, Ophuls fears that ecological scarcity will only intensify the competition among the separate states, thus leading to a "war of all against all."

In attempting to solve the global problem of the (now) faulty assumption of ecological abundance, Ophuls returns once again to the state. His recognition that the problem of the ecological commons is not confined to the separate territories of sovereign states does not lead him to a critique of state space as insufficient to house the flows of politics. The only solution Ophuls can see for a global problem is an extension of the reach of the institutions of the state— the problem does not lie with these institutions, but with the assumptions about the carrying capacity of the commons.

Ophuls's ecopolitical analysis contains no appreciation of the role of worldwide machines. In his chapters on political economy, Ophuls's main point is that liberal political theory, particularly as it has worked itself out in the United States, has attempted to keep politics and economics separate.[30] This has led to a free reign by the economic actors within the United States to overexploit our resources. Moreover, when he takes up a discussion of international political economy, it remains (as I have already mentioned) couched within standard international political theory states-as-actors terms.[31] The extent to which global political economy transverses sovereign state boundaries exceeding the limits states attempt to place on it receives no discussion by Ophuls. The worldwide machine of global capital remains safely contained within the space of the state.

The picture of nature provided by Ophuls also offers no significant break from standard modern Western thought on the subject. In fact, we are presented with almost a carbon copy of Marsh's vision of nature examined in chapter 1. Nature, Ophuls instructs, is "a complex organization made up of highly compatible parts; [that] in the absence of intervention, . . . runs as beautifully as a well-made watch (and never needs repairs)."[32] The similarities to Marsh's argument become all the more evident when the title of one section in Ophuls's chapter on the science of ecology ("Man the Breaker of Climaxes")[33] is compared to the original titled suggested by Marsh for his 1864 work ("Man the Disturber of Natural Harmonies").

Ophuls's politics offers a glimpse of what Marsh might have advocated, given the similarities in their conceptions of 'man' and nature. For both, 'man' is the disturber of nature's harmonies, harmonies that would sound forever, presumably, if it were not for 'man''s propensity to "tamper randomly with [this] delicate instrument."[34] Given that propensity, and given the apparent lack of self-restraint among 'men' to heed the necessities of ecology, the only hope lies with a sovereign authority with a sufficient amount of coercion to bring 'man''s actions into harmony with nature.

Ophuls shows no appreciation for the chance of nature in this work. It is interesting to note that the description of nature is the one area where Boyan significantly revises Ophuls's argument in *Ecology and the Politics of Scarcity Revisited*, presenting a picture of nature much closer to Botkin's than Marsh's. Boyan revises/updates Ophuls's descriptions of nature to accompany shifts in the field of ecology about the character of natural processes described in chapter 1. In Boyan's hands, natural processes take on a greater randomness. Gone are the descriptions of nature running like a well-made watch. Gone is the argument that if there were no *design* limits, there would be no system at all.[35] For Boyan, limits are a matter of necessity, elements of a system that change as the system changes due to external or internal influences.[36]

I do not mean to suggest that the politics expressed in *Ecology and the Politics of Scarcity* is no longer applicable given the shift in conceptions of how nature operates from the 1977 version to the 1992 version. As I argued in chapter 1 with regard to chaos theory, changing one's conception of natural processes does not necessarily alter one's ethics—nor, I would now add, one's politics. This differs, however, from the claims I made concerning the relationship of human beings to the earth and our ontological privileging of our position in that relationship. Some politics are unable to sustain themselves in a world where there is no necessity to humanity, where the earth was not created with us in mind. I will return to this claim.

BOOKCHIN AND THE PROBLEM OF HIERARCHY

> No ecological society, however communal or benign in its ideals, can ever remove the "goal" of dominating the natural world until it has radically eliminated the domination of human by human, or, in essence, the entire hierarchical structure within society in which the very notion of domination rests.[37]

Within O'Riordan's and Dobson's matrix, the work of Murray Bookchin occupies the opposite position from Ophuls. For Bookchin, the call for a restrengthened sovereign—albeit a democratic sovereign—even for ecological purposes, flies in the face of ecology's primary principles. A politics of ecology must instead begin with the recognition that "the attempt to dominate nature stems from the domination of human by human; [and] that . . . harmoniz[ing] our relationship with the natural world presupposes the harmonization of the social world."[38] Any attempt to formulate a politics of ecology without first addressing the problem of social domination is doomed. "Whatever power the State gains," Bookchin argues, "it always does so at the expense of popular power."[39]

Bookchin writes a thoroughly antistate ecopolitics. The state exists, according to Bookchin "only when coercion is *institutionalized* into a professional, systematic, and organized form of social control . . . with the backing of a monopoly of violence."[40] In order to rule over a great expanse of people and land, the state must institutionalize coercion. The institutionalization of coercion sets the state off from other communities; it is what makes the state "a realm of evil," and hence, turns "the 'art' of statecraft [into] essentially a realm of lesser or greater evils, not a realm of ethical right and wrong."[41] Environmentalists that occupy the halls of state power are necessarily implicated in this evil. They are locked into a deadly game of reformism, an "ongoing *process* of degeneration."[42] The only hope for an ecological politics, according to Bookchin, is to break from the state model.

What makes the state primarily anti-ecological is the presence of hierarchy. In Bookchin's eyes, hierarchy does not exist in nature.[43] Hierarchy implies an organized system of rule. One should not read into this claim about nature a rhizomatic interpretation. As I will argue, Bookchin's nature, like Ophuls's, is indeed quite far from the rhizomes of Deleuze and Guattari's nature, or the chaos of Nietzsche's. Nature does not lack organization, from Bookchin's standpoint, but rule. Hierarchy underlies all systems of *institutionalized* rule. Whether based on the rule of females by males, the young by the old, the poor by the rich, the governed by the governors, hierarchies create the necessary preconditions for statecraft.[44]

If a truly natural society, a truly ecological society, is to be constructed, the state must be eliminated. An ecologically informed anarchism becomes the only way, on Bookchin's reading, to harmonize human-human relations, and thus human-nature relations. Again, this should not be read as an engagement with the nomadic trajectory seen in Deleuze and Guattari's *A Thousand Plateaus*. While both Bookchin and Deleuze and Guattari are critiquing the state model, Bookchin is doing so by establishing his anarchism as "the absolute negation of the state."[45] Deleuze and Guattari's nomadism is not the negation of the state, not its bipolar opposite, not its antithesis; it is, however, different from the state. Moreover, Deleuze and Guattari do not offer the nomadic orientation as the answer to the problems of the state. Danger, they are well aware, lurks in a nomadic orientation, too.

In order to create an "ecologically sound society," Bookchin argues, the "decentralization of large cities into humanly scaled communities" is essential.[46] Cities and nation-states have served to atomize people. They have created the very self-interested selves Ophuls's sovereign would seek to constrain. Ironically, the very possibility for this humanly scaled existence seems to emerge from the economics of the self-interested self—bourgeois capitalism. "If it achieved nothing else," Bookchin acknowledges that bourgeois society

> revolutionized the means of production on a scale unprecedented in history. This technological revolution, culminating in cybernation, has created the objective, quantitative basis for a world without class rule, exploitation, toil or material want. The means now exist for the development of the rounded man, the total man, freed of guilt and the workings of authoritarian modes of training, and given over to desire and sensuous apprehension of the marvelous.[47]

In a passage reminiscent of Marx's vision of the postcommunist world, Bookchin argues that the freedom necessary for eco-anarchism to succeed depends on human beings being "free *concretely*: free from material want, from toil, from the burden of devoting the greater part of their time . . . to the struggle with necessity."[48] But Bookchin breaks from Marx for almost the opposite reason Ophuls breaks with Locke. "The Marxian critique is rooted . . . in the era of material want and relatively limited technological development."[49] Where Ophuls argues that liberalism began with assumptions of abundance and now faces a world of limitations and scarcity, Bookchin maintains that in the mid-nineteenth century capitalism did face conditions of scarcity and Marx reacted to this situation, but now the scarcity we face is imposed and is a necessary component of bourgeois state structures.[50]

The capitalist economy has created the technological development necessary to provide for the needs of all. The hierarchical structures of the state prevent this from occurring. The anarchism Bookchin appeals to would, he maintains, remove these restrictive and unjust hierarchies and liberate technology and humans to meet the needs of everyone. But Bookchin's humanly scaled eco-anarchism would, in Gustava Esteva's eyes, fail to see itself with sufficient depth. Noting the role that global capital has played in creating the conditions of real abundance (and enforced scarcity), Bookchin fails to consider the extent to which this worldwide machine will not be captured by his attempt to decentralize into humanly scaled communities. That Bookchin expects the abundance from global capital to be available even after he has dismantled it suggests a lack of appreciation for the complexities of global ecopolitics.

Where Ophuls's politics requires a government strong enough to check the overexploitative tendencies of individuals, Bookchin argues that the self-interested self that would require such a government is an aberration, an unnatural product of social hierarchy. Bookchin's humanly scaled decentralized political communities would allow the individual to shake off self-interest and environmentally destructive behavior, and attune to nature. The problem, as Bookchin understands it, is not that the individual, or humanity, is unfit for a politics of ecology and must be more effectively controlled or guided by the state, but that the individual, or humanity, has been denigrated within hierarchical societies.

Bookchin's appreciation of the production of this unnatural individual is not an appreciation for Foucault's notion of governmentality. Where Foucault rejects the notion of a "natural human" devoid of the productive aspects of societal apparatuses, Bookchin bases his eco-anarchism on a belief in the truly natural person.

"The individual is, indeed truly free," Bookchin contends, "and attains true individuality when he or she is guided by a rational, humane, high-minded notion of the social and communal good."[51] Humanity, for Bookchin, unlike Ophuls (or Marsh), is not the breaker of natural climaxes but "the potentiality for nature to become self-conscious and free":[52]

> The great achievements of human thought, art, science, and technology serve not only to monumentalize culture, *they serve to monumentalize natural evolution itself.* They provide heroic evidence that the human species is a warm-blooded, excitingly versatile, and keenly intelligent life-form—not a cold-blooded, genetically programmed, and mindless insect—that expresses *nature's* greatest powers of creativity.[53]

Thus the removal of unnatural hierarchical structures will allow for the teleological link between humans and nature to reestablish itself. What's more, natural evolution "is not a 'catch-as-catch' can phenomenon. It is marked by tendency, by direction, and, as far as human beings are concerned, by conscious purpose."[54]

Hierarchical systems of government threaten, rather than enhance, the process of natural evolution. "The issue . . . is not whether social evolution stands opposed to natural evolution. The issue is *how* social evolution can be situated in natural evolution and *why* it has been thrown . . . against natural evolution."[55] The "why" question has been answered through the presence of hierarchies. The "how" question is resolved through the creation of "a rational, ecologically oriented society."[56] Bookchin need not be concerned with the intemperate, the unecological, as Ophuls is. From Bookchin's perspective, these individuals are not natural. They will disappear when the artificial structures that produced them are destroyed. When this happens social evolution can once again tap into the "conscious purpose" of natural evolution through "a *free* society . . . conceived as a unity, a 'one' that is bathed in the light of reason and empathy."[57] Eco-anarchism is the only solution.

Bookchin's ecopolitics is based on a vision of nature as a glorious dialectical process, moving toward self-consciousness and freedom. Humanity, in Bookchin's eyes, is nature's greatest achievement, nature's hope to become self-conscious and free. This human, contrary to Marsh or Ophuls, is not destined to destroy nature's harmonies. Bookchin's human has the potential to attune to the wonderful telos of natural evolution.

Ophuls and Bookchin differ on their conceptions of human nature, the character of nature, the presence of scarcity, the role of state institutions in causing ecopolitical problems, and the role of the state in solving these problems. Where Ophuls speaks of a mechanistic nature with clocklike perfection, Bookchin speaks of an organic, dialectical nature. Where Ophuls presents humans as tamperers with the delicate machine of nature, Bookchin sees humans as (potentially) glorious examples of nature's teleology. Where Ophuls sees the ecopolitical problem as caused by a politics based on abundance and faced with scarcity, Bookchin sees the scarcity as enforced and holds that the preconditions for abundance do exist. Where Ophuls argues that given the absence of ecotemperance within people, government must enforce it from without, Bookchin argues that govern-ment is the root cause of our ecopolitical problems and that the re-moval of government and its hierarchical structures will allow for the truly natural, ecological person to (re)emerge. The two appear to be, on many levels, at opposite ends of the ecopolitical spectrum.

Despite these differences, when viewed from a (post-)Nietzschean perspective, Ophuls and Bookchin appear as very similar thinkers. Both present natures with particular designs, natures that evidence qualities of being created by an intelligent being. "Natural history," to repeat Bookchin's claim, "is not a 'catch-as-catch-can' phenome-non. It is marked by tendency, by direction, and, as far as human be-ings are concerned, by *conscious purpose*."[58] For Ophuls, nature "runs as beautifully as a *well-made* watch."[59]

Furthermore, Ophuls's and Bookchin's theories are easily placed onto O'Riordan's and Dobson's matrix of ecopolitical thought due to their similar orientations to state and sovereignty. Although one endorses the state as the appropriate place for ecopolitics and the other holds the state to be the absolute negation of an appropriately ecological politics, both reify the state as the locus of political activ-ity—for good or bad. Neither appreciates the difference of political space that exceeds the sovereign territorial state model of political space.

A CYBORG ECOPOLITICS

A cyborg is a cybernetic organism, a hybrid of machine and or-ganism, a creature of social reality as well as a creature of fiction.[60]

At the time of Marsh and Nietzsche, machines were "not self-moving, self-designing, autonomous . . . Late twentieth-century machines," however, "have made thoroughly ambiguous the difference between natural and artificial, mind and body, self-developing and externally designed, and many other distinctions that used to apply to organ-isms and machines."[61] The differences between the competing ideals of machine and organism are thoroughly confused in the late twenti-eth century. The border between the two has been broken down through the creation of virtually organic machines and mechanical organisms. The cyborg is a product of this border confusion. It is a creature of both science fiction and scientific reality. Cyborgs are found in the worlds of sci-fi literature and movies, medicine, the mil-

itary, industry . . . For Haraway, the cyborg is the quintessential figure of late twentieth-century social and political life. The cyborg is a
border creature.

The cyborg "gives us our politics," Haraway suggests.[62] On one
level, Haraway can be read as suggesting that contemporary politics
exists on border lines. Not just the border lines between machines
and organisms, or humans and nature, but the border lines that
divide sovereign territorial states, or set apart public spheres from
private spheres. The cyborg exists on the border line between the
United States of America and Japan, where "Japanese" car companies manufacture cars in the United States, and "American" cars are
made in Japan. The cyborg exists on the border line between the
United States and Mexico, where cross-border corporations, pollution, illegal immigrants, and an increasing "feminization of work"[63]
disrupt the certainty of what counts as American and what counts as
Mexican, what counts as the public (male) sphere and what counts
as the private (female) sphere.[64]

Cyborg politics also problematizes the model of the body politic,
long thought of through an analogy with the human body where the
(sovereign) head rules over the rest of the (societal) body. In "Biopolitics and Postmodern Bodies," Haraway explores the implications of such an analogy in the face of late twentieth-century medical
discourse:

> The hierarchical body of old has given way to a network-body of
> truly amazing complexity and specificity. The immune system is
> everywhere and nowhere. Its specificities are indefinite if not infinite,
> and they arise randomly; yet these extraordinary variations are the
> critical means of maintaining individual bodily coherence.[65]

The organization of both the human body and the body politic has
shifted. The implications of this shift for political theory are profound. Politics can no longer be thought of as centered in a sovereign
figure. Can we easily locate the sovereign head of contemporary politics from which all organization emerges? Does a (conscious) head
exist that gives all the orders? Or is politics, like the immune system,
everywhere and nowhere, involving not just the state but worldwide
machines and local mechanisms, governmentality and disciplinary
power? "We are living through a movement from an organic, industrial society to a polymorphous, information system," Haraway con-

tends, ". . . from comfortable old hierarchical dominations to scary new networks . . . of domination."[66]

From an economic perspective, we could talk about this movement in terms of a shift from Fordism or industrial capitalism to flexible accumulation or finance capitalism. Finance capitalism, an information system, has increased the difficulty for the state to control capital. It has problematized the sovereign-head model of politics. Finance capital exists along cyborg networks, and is not isolated to particular sovereign territories or even traditional conceptions of time and space. "Much of the flux, instability and gyrating [of current global markets] can be directly attributable," geographer and social theorist David Harvey argues, "to this enhanced capacity to switch capital flows around in ways that seem almost oblivious of the constraints of time and space that normally pin down material activities of production and consumption."[67] The old model of the body politic is inadequate in the face of the networks of global capital—to name only one way in which politics problematizes the sovereign-head model for the body politic.

Haraway appropriates the figure of the cyborg in an attempt to come to terms with the scary new networks of this border line politics, to provide an image with which to envision contemporary social reality. Like the second "Terminator" in the movie *Terminator Two* (the one not played by Arnold Schwarzenegger), there is no central command "cell" in contemporary politics. The "cells" of political activity are interconnected, but there is no one "cell" on which all the other "cells" depend.

Haraway does not look to the cyborg as a salvational image. She appropriates the image of the cyborg in an attempt to "build an ironic political myth," or perhaps better put, a "blasphemous" one.[68] "The main trouble with cyborgs, of course," Haraway admits, "is that they are the illegitimate offspring of militarism and patriarchal capitalism, not to mention state socialism."[69] As a potential spokesperson for an ecological politics, the cyborg does not seem to carry the proper credentials. Militarism, patriarchal capitalism, and state socialism are arguably the three largest destroyers of life on earth. Put in Nietzschean terms, the 'men' of each of these three "isms" may well be manifestations of "the ugliest man," that murderer of God who would take God's place and assume dominion over the earth. "From one perspective, a cyborg world is about the final imposition of a grid of control on the planet, about the final abstraction embodied in a Star Wars apocalypse waged in the name of defence, about the final appropriation of women's bodies in a masculinist orgy of war."[70] Haraway is profoundly aware of the destructive capacity of the cyborg world.

But cyborgs are the *illegitimate* offspring of the 'men' who may bring this destruction. This illegitimacy may be their saving grace, if you will. As Haraway is quick to point out, "illegitimate offspring are often exceedingly unfaithful to their origins. Their fathers, after all, are inessential."[71] While highlighting the dangers, Haraway looks to the cyborg for possibilities. She suggests potentialities for a cyborg politics to be a radical politics, an ecological politics. She attempts to write a(n eco) politics that engages the network world she describes.

On one level, a cyborg politics breaks from the sovereign territorial state structure of politics. O'Riordan's and Dobson's matrix is not sufficient in this cyborg world. The sovereignty model is insufficient not just because the physical effects of our decisions spill across national frontiers, for as my reading of Ophuls suggests, the boundaries of state structures can be extended. The sovereign model is insufficient because the relations of capital that are a part of these decisions are not confined to the space of the sovereign territorial state. Haraway establishes a political problematic that exceeds this sovereignty/territory frame. Unlike Bookchin's eco-anarchism, her cyborgism is not intended as the negation of the state. More in keeping with Deleuze and Guattari, Haraway's works suggest a different space of (eco)politics.

Cyborg politics problematizes the space of the state. The state becomes difficult to locate when cyborg aspects of politics are taken seriously:

> Continued erosion of welfare state; decentralizations with increased surveillance and control; citizenship by telematics; imperialism and political power broadly in the form of information rich/information poor differentiation; increased high-tech militarization increasingly opposed by many social groups; reduction of civil service jobs as a result of the growing capital intensification of office work, with implications for occupational mobility for women of colour; growing privatization of material and ideological life and culture; close integration of privatization and militarization, the high-tech forms of bourgeois capitalist personal and public life; invisibility of different

social groups to each other, linked to psychological mechanisms of belief in abstract enemies.[72]

The state is not absent in this cyborg world, but it does exhibit a different character. Public and private spaces are blurred. Where political theorists have valiantly struggled to keep the space of the state, and hence public space, pure of all "nonstate" or private contaminants, Haraway elucidates numerous levels from which attempts to cast politics as solely state activity are problematic. As with Foucault's discussion of disciplinary power, Haraway's discussion of cyborg aspects of politics bring out multiple ways in which politics is not reducible to the (public) space of the sovereign head. The cyborganizations of offices and factories, presumably operating within the private sphere, do not exist independently of the professed (state) political realm. They have had profound impacts on the activity within the public space of state capitals, as well as the private space of households. Haraway's use of the cyborg—a borderline creature that has broken down the barrier between machine and organism, problematizing what it means to be either—as an image for politics is intended to break down political boundaries as well, problematizing the concepts we use to think about politics.

Despite this emphasis on border lines and boundary confusion, Haraway contends that she is not erasing location. "I am arguing for politics and epistemologies of location, positioning, and situating," she insists.[73] In this sense she is perhaps closer to Bookchin than to Ophuls. But where Bookchin seeks to write a politics of locality and decentralization in an attempt to flee from the evils of nation-state or global politics, Haraway writes a politics of positioning to demonstrate the highly local effects of worldwide networks of domination. She is operating within a different political problematic: one in which the state has already lost is place of primacy; one in which politics already operates in nonstate manners and the destruction of the earth has not dissipated. As with Esteva, Haraway sees a necessity to highlight location in the light of global processes and global systems in the light of local effects.

"Third World" debt and the capital transfers that accompany it are bound up in Brazil, for instance, with debates over deforestation, extractive reserves, the debt peonage of rubber tappers, the "sovereignty" of indigenous peoples, the preservation of the jaguar, and so on.[74] What decisions are made regarding the finance capital at play here will most certainly change the way many people in the Amazon jungle (and perhaps around the globe) live their lives. It is crucial not to forget the local aspects of global networks. For those of us interested in "saving the Amazon," our struggle (as "First World" environmentalists) cannot be divorced from the lives of the human forest dwellers, anymore than it can ignore how the political economy of the state of Brazil and the multinational corporations and lending institutions that permeate the Amazon intersect our lives. Again, it is difficult to locate the sovereign center of this politics; it is difficult to draw boundaries about this territory. It is incredibly diffuse, while being particularly local.

While insisting on a politics of positioning, Haraway, once again, highlights potential dangers. There is a danger of "authenticity" that involves granting speakers *sole* voice on an issue because of their position. Does the president of Brazil have the final say in the debate over what to do with the Amazon rain forest due to his position as not just a Brazilian but the "legitimate ruler" of that territory? "I don't know," Haraway remarked in an interview, "maybe one should occasionally use discourses of authenticity. But I tend to be wary of them because they tend to work as taxonomies, as zoos— each to her or his own authentic position. And if you don't have one you have no say . . . each to his authentic place is apartheid."[75]

Discourses of authenticity tend to privilege a certain politics of place that coincides with a politics of sovereign territory. They work to dismiss claims from groups or individuals who appear to have no place and hence operate to force everyone into a particular place. Witness sovereign state readings of "Indians" on both accounts. "Indians" had no claims to lands in the American continents because they occupied no particular place; and furthermore, if they were to become "civilized" and hence citizens of states, they had to be taught to settle permanently.

Haraway's claim that the cyborg "gives us our politics"[76] subverts this politics of authenticity. Cyborgs have no "authentic place," whether that be politically, socially, sexually, ethnically, racially, or linguistically[77]—which does not mean all cyborgs are the same, just that the places they occupy are "already shot through with difference."[78] In other words, to push Haraway's argument, a recognition that "we" are cyborgs is a recognition that we do not occupy sovereign locations, or categories in the sense that we are solid identities. It is a recognition that "our" identities are permeated with difference. Applied to the discussion of sovereign territory, the image of the cyborg explodes the myth of the unified state. In a cyborg world of networks of domination, every sovereign state territory is always already filled with difference.

Although Brazilian president Sarney made an important point to "First World" politicians and environmentalists when he issued his "Our Nature" proclamation, he cannot unproblematically be argued to have spoken with the authentic voice of Brazil. The "our" in his edict is far from unitary or all-encompassing. Who does Sarney's "our" include? The numerous indigenous tribes that live throughout the territory of Brazil? The thousands of children living in the streets of Brazil's major cities? Rubber tappers? The military? Transnational corporations and lending institutions that are helping Brazil manage "its" nature? Moreover, the space that Sarney's "our" attempts to encircle, as I have been arguing, is far from unproblematic as well.

Discourses of authenticity also contain a "danger of romanticizing and/or appropriating the vision of the less powerful while claiming to see from their positions."[79] "First World" environmentalists run this risk when they attempt to speak for, or in the interest of, indigenous or "Third World" peoples. Susanna Hecht and Alexander Cockburn, coauthors of *The Fate of the Forest: Developers, Destroyers, and Defenders of the Amazon*, point out how "First World" environmental organizations frequently sanitize local "Third World" movements of their local politics in order to present their struggles in a palatable form to other "First Worlders."[80]

The assassination of rubber tapper labor leader Chico Mendes, for example, has been utilized by numerous "First World" environmental organizations to garner support and funds for the fight to save the Brazilian rain forests. Mendes's death has become a symbol of the death of the tropical rain forest. But from a local perspective, Mendes's death was less about a global struggle to save the rain forests than a local struggle between disempowered rubber tappers and powerful cattle ranchers.[81] Fighting to preserve a way of life and the environment that it depends on, rubber tappers across the Amazon are engaged in battles with other economic interests that want to drive them from the jungle so that it may be transformed into cattle ranches or mines. Mendes's death has become a global symbol of the dying rain forest, yet it is also intensely about local political economy. Again, location does not disappear in a cyborg politics, but it cannot be thought of independently from global processes. The death of a labor leader involved in a struggle over land use in the Amazon tropical rain forests cannot be divorced from the global flows of capital also involved in struggles over land use in the same forests—and vice versa. Ecopolitics is not about one perspective.

"The political struggle," Haraway contends, "is to see from [multiple] perspectives at once [given the dangers pointed to earlier] because each reveals both dominations and possibilities unimaginable from . . . other vantage point[s]."[82] To exist in this cyborg world requires multiple vision. Not a vision from everywhere, for that is a "god-trick,"[83] but a vision from somewhere with multiple perspectives. To exist in this cyborg world requires the ability to see not only the modes of domination operating in this world, but the possibilities for resistance as well.

Not born innocent, the cyborg is a perfect vehicle through which to obtain this vision. A bastard child of militarism, patriarchal capitalism, and state socialism, the cyborg is deeply aware of the dangers of these network modes of domination. But it is also aware of the ecopolitical possibilities that reside in these networks: alliances between indigenous inhabitants of rain forests and "First World" environmental organizations; alliances between government bureaucrats and Earth First!-ers; full-page ads in the *New York Times* or the *Wall Street Journal*; boycotts that situate themselves along the network lines of global capital, rather than orienting themselves exclusively around sovereign territories, and so on.

Haraway's visions of nature and humanity harken back to descriptions of chaos theorists and Nietzsche addressed in chapter 1. Hence, the ethics that accompanies her thought parallels in many ways the Nietzschean ethic I attempted to articulate. Unlike Ophuls or Bookchin, nature for Haraway is neither a well-made watch nor a teleological process. It evidences an agency that confounds the mechanistic model. "The Coyote or Trickster, embodied in American Southwest Indian accounts, suggests our situation . . . knowing all the while we will be hoodwinked."[84] Nature as chaos, perhaps? It also fails to provide "a source of insight and promise of innocence"[85] necessary for teleological organicism. Nature, for Haraway, is a source of contingency and diversity. It is under no obligation to turn towards us a legible face. In this sense, Haraway's vision of nature confounds theories that place human beings in an ontologically privileged position.

The cyborg is a creature of science (fiction). And according to contemporary science, Haraway contends, human beings have lost their privileged place in the grand scheme of things. "Biology and evolutionary theory over the last two centuries have . . . reduced the line between humans and animals to a faint trace."[86] The list of characteristics that have served to separate humans from the rest of the animal world has been severely problematized. Language, problem solving, social organization, and so on can all no longer be considered the specific domain of the human being. And following from this, it has become increasingly difficult to argue that the world was created specifically with us in mind. The cyborg political theory offered by Haraway exists beyond the shadow of God in these respects.

From her descriptions of nature, to her subversion of the human conceit that the world was created for us, Haraway's cyborgism contains crucial elements at work in my Nietzschean eco-ethic. Unlike Ophuls and Bookchin, Haraway's ecopolitical theory is not built upon a utilitarian human-earth relationship. For Haraway, nature is Coyote, the Trickster. It is neither property to be appropriated by our labors nor a vast resource available to provide for a socially harmonious life. The cyborg must have a different orientation to the world. As with Nietzsche, Haraway seems to be advocating a reverence for the difference, the chaos of nature.

5

Brazil of the North

It was no coincidence that the 1992 United Nations Conference on Environment and Development (UNCED), or "Earth Summit," took place in Brazil. Home to the rapidly disappearing Amazon rain forest, Brazil has come to symbolize the global environmental crisis. This symbolic status of the Amazon and Brazil contributes to the assumption that the global environmental crisis is largely about activities in lands south of the equator, or at least not a great distance north of the equator. Or maybe it is this assumption that has made Brazil the symbol of the global environmental crisis. Either way, it is important to realize that the destruction of the global environment is not centered in the south, that, for instance, the tropical rain forests of Brazil (and elsewhere in the south) are not the only rain forests in the world at risk. Rain forests in the northern hemisphere are also being cut down at an alarming rate.

Approximately one-half of all remaining temperate rain forests on the earth lie along the west coast of North America (even though only about 8 percent of what was there a century ago remains). The west coast of Vancouver Island in British Columbia, Canada, is home to some of the largest remaining contiguous uncut areas of this type of forest. Yet only one-third of the old growth on Vancouver Island remains and 90 percent of that is scheduled to be cut down in the next few years. Dubbed "the Brazil of the North" by various environmental organizations because of the logging practices in operation there, the temperate rain forests of Vancouver Island have recently entered the global spotlight.

Temperate rain forests are dominated by various species of hemlock, cedar, fir, and spruce trees that commonly live as long as eight hundred years (some well over a thousand) and grow to heights of ninety-five meters (or over three hundred feet). Temperate rain forests contain a higher biomass per acre than any other forest type in the world. Upwards of 4,500 different species, ranging from bald eagles and marbled murrelets, to salmon and orcas, to wolves and black bears, to salamanders, mosses, lichen, and so forth, have been discovered in one 260,000-hectare area of Vancouver Island known as Clayoquot Sound.

As with tropical rain forests, much of the biodiversity of the temperate rain forest grows not out of the soil, but out of the decaying trees that occupy the forest floor. Long thought of as waste or fire hazard, deadwood is now recognized by many forest ecologists as crucial to a forest's life. Not only can fallen trees protect the soil from torrential rainstorms, but they provide habitat for hundreds of animal species, not to mention their function as nurseries for new plant life.

Additionally, the temperate rain forests of North America are largely coastal and are inextricably linked to the ocean ecosystems that border these forests. The logging of watersheds that house streams and rivers that eventually flow into the ocean has had disastrous effects on salmon populations. In consequence, this logging has had a significant impact on a variety of ocean life that depends in one way or another on the salmon. In other words, the ecosystem of the coastal temperate rain forest does not end at the shoreline. Of the ninety watersheds over five thousand hectares in size on Vancouver Island, only five remain unlogged. Three of those five untouched watersheds are in Clayoquot Sound.

Vancouver Island has received global attention not just because ancient trees are being cut there, but because of the manner in which they are being cut. As with Brazil, the principal method of logging on Vancouver Island is clear-cutting.[1] The process of clear-cutting involves cutting and removing all trees and plant life in a particular area, not just the economically desired ones. While clear-cutting is argued to be the most economically efficient manner of extracting the valued trees, it is far from ecologically efficient. Aside from removing the trees themselves, clear-cutting also removes the deadwood that covers the forest floor, thus depleting the area of available habitat for a wide variety of temperate rain forest species. Clear-cutting in this manner also exposes the soil to torrential downpours, often resulting in large mudslides and massive erosion. Furthermore, it often leaves "islands" of trees too small to sustain themselves and the wildlife that resides within them. The British Columbia government's own independent scientific panel has condemned clear-cutting in Clayoquot Sound for not only ecological, but cultural and future-oriented, considerations.[2]

Vancouver Island provides an excellent "case study" for the discussion of ecopolitical space I have undertaken in this work. The following discussion of the politics of deforestation in Clayoquot Sound in particular and British Columbia in general will help draw out the insufficiency of the various discourses of sovereign territorial political space I have examined in earlier chapters, from President Sarney's "Our Nature" proclamation to the World Commission on Environment and Development's recognition that the effects of "our" decisions spill across national boundaries. My claim is not that the ecopolitics of deforestation in Clayoquot Sound (or elsewhere) is not about sovereign territorial politics as domestic or international policy. My claim is that the problem of ecopolitics cannot be reduced to either domestic or international policy—that we need to think the space of ecopolitics beyond sovereign territory.

First of all, ecopolitical issues exceed sovereign territory as they slice across the pathetically porous geopolitical boundaries that constitute British Columbias or Canadas occupying what I described in chapter 2 as rhizomatic or smooth space. Second, exploring the space of ecopolitics takes us into the realm of "governmental" practices (chapter 3) that construct both the territories and the populations available for sovereignty. Both are evident in Clayoquot Sound. Finally, an investigation of the ecopolitics of deforestation on Vancouver Island provides one more opportunity to flesh out my Nietzschean eco-ethic from chapter 1 and to solidify connections between it and the issues raised in the other chapters.

SOVEREIGNTY

In the summer of 1993 the Australian-based rock band Midnight Oil participated in a blockade of a logging road in a clear-cut area in

Clayoquot Sound. A common question addressed to members of the band while in Clayoquot Sound by the local media was, Why are five musicians from Australia participating in a local political struggle in faraway Clayoquot Sound? Why aren't you back in Australia pursuing environmental issues there? The band's ability to speak on logging in Clayoquot Sound was challenged by the location of their homes—that is, not Clayoquot Sound, or Vancouver Island, or even British Columbia. From the perspective of their critics, Midnight Oil lacked the authenticity to comment on logging practices or participate in a struggle that was going on halfway around the world from where they lived.

The position of these critics of Midnight Oil's presence in Clayoquot Sound is not dissimilar to the position of former Brazilian president Sarney when he issued his "Our Nature" plan for the Amazon rain forest (see Introduction). In both cases, the boundaries of sovereignty are being policed by those who insist that in order to have any type of say one must be situated in that place. This is a discourse of authenticity. In other words, one must be a Canadian, or even a British Columbian, to comment on the issue of deforestation in this area. In keeping with Haraway's comments on the discourse of authenticity presented in chapter 4, I am not saying that location is not important, or even crucial, to ecopolitics; instead, I am suggesting that territorial discourses of authenticity operate to silence the fact that ecopolitics is not territorially confined. Midnight Oil's presence on Canadian soil draws out the permeability of the sovereign territorial boundaries that operate to create this authentic position.

The voice of Midnight Oil's critics is a Waltzian voice (chapter 2) that assumes a legitimacy of rule over a particular territory. The internal difference that constitutes British Columbian or Canadian politics is smoothed over perhaps through the claim of majority rule, or at least through the presence of a geopolitical border separating British Columbia/Canada from other sovereign territories. But this position elides the nonsovereign territorial nature of ecopolitics, the extent to which the logging of the temperate rain forests on Vancouver Island is not confined to the geopolitical boundaries of this sovereign territory but is bound up with a global market for wood products. This position also erases the internal difference of British Columbia/Canada brought out by the presence of non-Canadian

Native American nations that still claim "sovereignty" over much of Vancouver Island, not to mention British Columbia or Canada.

Midnight Oil's presence in Clayoquot Sound helps accent the spatial difference of ecopolitics. On the one hand, the band was there to lend support to hundreds of Clayoquot Sounders fighting an intensely local fight to preserve the last of the ancient temperate rain forests. On the other hand, the band's presence in Clayoquot Sound helped extend the space of this fight in a rhizomatic manner across the globe through the networks of the global media—networks that parallel, in many ways, the production, trade, and consumption of Clayoquot Sound wood products. Although the politics of the forests of Clayoquot Sound cannot be divorced from their locality, they cannot be reduced to it. The challenge to Midnight Oil's presence in Clayoquot Sound operates to deny the extent to which the logging of ancient trees in one part of the world takes places because of trade and consumption patterns in other parts of the world. It also operates to deny the extent to which the ethical issue of logging these forests is not contained by the assertion of geopolitical boundaries. To isolate the issue to the geopolitical boundaries of British Columbia/Canada, to reduce it to an issue of domestic or even international policy, hides much of what is occurring there. Midnight Oil's presence in Clayoquot Sound helped to (re)focus attention beyond the space of sovereign territory.

MacMillan Bloedel, Inc., the main logging company operating on Vancouver Island, "had seen its worst nightmare come true," reported the *Vancouver Sun* on 12 November 1993. "The remote temperate rain forest on the West Coast of Vancouver Island was en route to becoming the global icon of the conservation movement."[3] In October and November of 1993, protests and demonstrations highlighting the deforestation of Clayoquot Sound took place in Hamburg, Bonn, Munich, Vienna, London, Sydney, and Tokyo. According to Dennis Fitzgerald, head of environmental communications for MacMillan Bloedel (MB), MB was ill equipped to deal with such a politicization and globalization of its logging activities: "We know how to manufacture pulp and paper, but we don't know how to manufacture public opinion on an international scale."[4] MB, a multibillion-dollar global corporation, is ill equipped to fight battles on a global scale, we are told. Or, put another way, MB is said to be adept at activities in the private sphere (economic activities), whereas manufacturing public opinion, presumably a public sphere activity, is said to be outside its range. By sidestepping the role of marketing and advertising as ways in which MB shapes public opinion, the claim is an interesting one.

The issue of MB's activities would be easier for MB to handle, it seems, if the boundaries of politics were clearly maintained: if economic activities could be contained within the private sphere and not become "politicized"; and (should that fail) if activities within the sovereign territory of Canada could remain solely a part of Canadian politics, where MB seems to be more adept at "manufacturing public opinion." What threatens MB, and corporations like it, is that the environmental fight against it will escape the space of the sovereign territorial state and confront it in different spaces. But those different spaces are the spaces of ecopolitics. The spaces, moreover, in which MB itself operates. Rhizomatic spaces. Smooth spaces . . .

Environmental groups, ranging from the Clayoquot Sound-based Friends of Clayoquot Sound (FOCS) to the global Greenpeace, took what was supposed to be a British Columbia political issue (the clear-cutting of ancient temperate rain forests in Clayoquot Sound) to the global marketplace. According to Valerie Langer, a member of FOCS, their group realized quickly that the deforestation in Clayoqout Sound is not just a local or provincial issue between citizens and government, but is deeply involved with where British Columbia wood products are purchased throughout the world.[5] Moreover, FOCS realized it might stand little chance of influencing the British Columbia government after it became the largest single shareholder of MB stock in February 1993, just months before it issued its "land use" plan for Vancouver Island. Recognizing that the politics of deforestation in Clayoquot Sound is not isolated to the space of British Columbia's or Canada's capitol buildings, FOCS took its battle to other spaces.

Canada is the world's largest exporter of wood pulp and newsprint. Trees from British Columbia produce 29 percent of that exported pulp and 16 percent of that exported paper. As of March 1994, Clayoquot Sound trees were being consumed in the United States in the form of Pacific Bell phone directories, Yellow Pages, the *Wall Street Journal*, *USA Today*, the *New York Times*, the *Seattle Times*, the *San Jose Mercury News*, the *Santa Rosa Press Democrat*, among others.[6] Aside from the United States, Japan, Germany, and the United Kingdom are the top three importers of British Columbia forest products and MB is the principal forest products company on Vancouver Island.[7] So FOCS, with the help of Greenpeace, Friends of the Earth, and the Women's Environmental Network, among other environmental groups, took its argument to Europe and the United States in 1993, attempting to convince individuals, groups, governments, and corporations to boycott British Columbia wood products. Their efforts have led (so far) to the cancellation of MacMillan Bloedel contracts by Scott Paper and Kimberly-Clark (the makers of Kleenex), both based in England, and to a pledge from the German magazine *Stern* to search for a clear-cut-free paper source.[8]

In addition to the activities of these environmental groups, an independent U.S.-based team of investigative reporters for the television program *The Crusaders* took the Clayoqout Sound issue to the doors of the *New York Times*. Pointing to a *Times* editorial that claimed "There's no reason . . . why the richest country on earth cannot save its old-growth forests,"[9] *The Crusaders* felt that the *Times* ought to heed its own editorials and not print its words on paper made from old-growth forests. While putting together the segment that would air in April 1994, the reporters spoke with Steve Golden, vice president of forest products for the *Times*, and were told that the editorial policy and the business policy of the *Times* are two different things.[10] This statement is one more demonstration of the ways in which the space of ecopolitics gets compartmentalized, this time through a division between editorial policy (politics) and business policy (economics)—a position not too different from that of MB on the difference between manufacturing pulp and manufacturing public opinion. Despite the dismissal by Steve Golden of the apparent contradiction between the *Times*'s editorial and business policies, by the time *The Crusaders* aired the segment, the *Times* had canceled its Clayoquot Sound contract with MB only to sign a new MB contract from another mill, outside of Clayoquot Sound, but still in British Columbia and still making paper by clear-cutting ancient temperate rain forests.[11]

Both FOCS and *The Crusaders* activities demonstrate some of the ways in which the politics of Clayoqout Sound logging exceeds the sovereign territorial boundaries of Canada, both by extending beyond the geopolitical boundaries of the state and by extending beyond the "public sphere" of state politics. Both the British Columbia government and MB are aware of the space in which this ecopolitical battle is being waged. The British Columbia government allocated $7 million (Canada) in 1994 to an international public relations campaign spearheaded by the logging industry to counter the effects environmental groups have had.[12] Even British Columbia prime minister Mike Harcourt has traveled abroad to lend his support to this counteroffensive, perhaps further blurring the distinction between public and private.

The fact that the environmental movement is a "movement without borders" has been frustrating for the British Columbia government, according to Langer.[13] Unable to contain the issue within its sovereign territorial borders, it has been forced to engage a battle beyond its jurisdiction. As with the Greenpeace boats that tried to prevent nuclear weapons testing by France in the South Pacific (chapter 3), FOCS activities pursue the space of ecopolitics beyond the striated space of the state. Even for some presumed allies this activity proves troublesome. According to Langer, one member of the ministry of the environment in Germany expressed exasperation that groups like FOCS did not pursue "proper" international political channels.[14] But groups like FOCS recognize that ecopolitics does not exist solely within the space of sovereign territory, whether that be in the form of parliamentary walls or proper international political channels.

The challenge to the politics of sovereignty presented by the ecopolitics of deforestation does not come just from individuals, groups, or corporations taking the issue beyond territorial borders; sovereignty is challenged from within its geographical territory as well. A bill that was proposed in the European Parliament in the summer of 1993 urging member states to impose a moratorium on all wood fiber products from British Columbia listed as its main reasons for the moratorium not just the logging practices in British Columbia, but also the issue of "native land claims."[15] The question of Canadian sovereignty over many lands now inhabited and once inhabited by the "First Nations" (inhabitants of what is now Canada whose ancestors predate the European "discovery" of the Western continent in the fifteenth century) has yet to be resolved.

Members of "First Nations" claim that Clayoquot Sound (as well as up to 80 percent of what is now British Columbia) was never ceded to the Canadian government in a treaty, nor lost in a battle, and hence does not rightfully belong to the British Columbia government.[16] Nonetheless, with the question of who holds rightful title to these lands still unanswered, the British Columbia government continues to grant logging licenses to strip these lands of their ancient forests. Said Andrew Peter, aboriginal affairs minister of British Columbia, "We are not going to shut down all industrial activity because we have not yet resolved the very important (land-claims issue)."[17] But should the land claims issue be decided in favor of the "First Nations," attempts are already being made to make sure members of "First Nations" take up the banner of "industrial forestry."

GOVERNMENTALITY

The problem of ecopolitics is not subsumed by the preceding discussion of sovereignty. Ecopolitics is also about government, or governmentality. Government, Rousseau argued in the mid-eighteenth century, is about "everything required by the locality, the climate, soil, moral customs, neighborhood, and all the particular relationships of the people . . . an infinity of details, of policy and *economy*."[18] Furthermore, the problem of government is about creating a population with the necessary characteristics to interact with particular lands in order to provide for the common good—which has now become global. Populations, if we recall Rousseau's discussion from chapter 3, do not come equipped to interact with lands to provide for the needs of the modern state. Moreover, lands do not come readily equipped to release the necessary products for populations, or opulent states. The practice of government, or governmentality, is about transforming both populations and lands to meet a desired end.

The problem of ecopolitics cannot be summed up by the World Commission on Environment and Development's claim that the effects of decisions in one sovereign territory have consequences in other sovereign territories. The problem of ecopolitics cannot be reduced to a problem of geopolitical boundaries. The problem of ecopolitics is very much a problem of who we, as populations inhabiting political communities, are and how we relate to the earth. The problem of ecopolitics, in other words, is about governmentality.

For members of "First Nations," the issue of land claims is not summed up by the logging battles throughout Vancouver Island and Canada. But for many non-"First Nations" inhabitants of British Columbia, especially those concerned with maintaining the logging industry, the issue of land claims cannot be divorced from the logging issue. "If First Nations obtain large, long-term forest tenures as part of land claims settlements, forest management skills will be essential," maintains the British Columbia Task Force on Native Forestry.[19] "First Nations," ancestors of peoples who have lived among these ancient rain forests for thousands of years, since before there was a British Columbia government or a logging industry, must, according to this task force, be taught forest management skills in case they should regain control over forested land. The governmentality witnessed in the eighteenth-century Portuguese American directorate lives on. The logging industry depends on it.

Although schools were set up in British Columbia over a hundred years ago through the Indian Act of 1876 to teach "Native" children how to become productive members of society, members of "First Nations" between the ages of twenty and forty still face a 50 percent unemployment rate. This unemployment rate will take on a new urgency if "First Nations" get control of British Columbia forests. The large numbers of unemployed "Natives" might not be sufficiently convinced of what their role as owners of vast tracts of forested land should be, or of the necessity of a vibrant logging industry for both the national and global economies. They might choose other ways to interact with these lands. The recommendations of the Task Force on Native Forestry are designed to create the kinds of people who will see the forest as first and foremost an economic resource. Moreover, these recommendations would see to it that the territory of British Columbia would continue to bring forth the products essential to the current provincial economy, products, furthermore, that are now deeply linked to a global economy. The recommendations of this task force are examples of a discourse of governmentality.

A priority for this task force is to see to it that any resolution of the land claims issue by the provincial and federal governments will provide for the stability of the "forest industry." And one primary manner of stabilizing the "forest industry" should "First Nations" receive a favorable settlement is to get "First Nations" members more involved in the "forestry industry" now, specifically through "cooperative management agreements and joint ventures."[20] Because more than 99 percent of available forest license land lies in the hands of non-"First Nations" companies, it would be difficult to get "First Nations" started on their own logging operations. Instead, ways need to be found to encourage members of "First Nations" to seek jobs in silviculture, where they would be engaged in activities like tree planting, spacing, pruning, and fertilizing—"natural occupations," according to this task force, for "Native people."[21]

The claim that there are "natural occupations" for "Native people" raises a number of questions: What is a "natural occupation"? What makes one occupation "natural" for a particular people? What occupations are "natural" for non-"Native people"? Are there "unnatural occupations"? From a Foucauldian perspective, this invocation of "natural" operates as a mask for the productive effects of disciplinary power. Just as women are not "naturally" predisposed to cook meals and do laundry, but are constructed into objects designed to engage in these activities by a host of disciplinary apparatuses, "Native people" are not "naturally" predisposed to tree planting, pruning, and so on. They are, however, being constructed into objects for that purpose. Just as the "Indians" of Portuguese America in the eighteenth century were to be made into productive members of society through their labors in the Amazon rain forests, twentieth-century "Native people" of British Columbia are to be made into productive members of the state and global economies through their labors in these temperate rain forests.

Citizens, or even "economically rational" individuals, are not given, but are created. The practice of government/ality, according to Rousseau and Foucault, is about customs, habits, opinions, and so on. Populations are not given but must be constructed. Where theorists like Bull (chapter 3) posit a given population across time and culture ready to take on the framework of sovereignty, Rousseau's and Foucault's investigations of government/ality challenge Bull's unproblematic assertion of population. And because sovereignty is presented as a system of rule over a given population, the problematizing of population, through a discourse of governmentality, is simultaneously a problematization of sovereignty. As long as "First Nations" are isolated to less than one-half of 1 percent of the land in Clayoquot Sound,[22] their characteristics can largely be ignored. They are no threat to the economy or sovereignty of British Columbia. But should they gain control over large areas of British Columbia territory, they must become part of the population of British Columbia, not just for economic reasons, but for sovereignty reasons as well. And this requires processes of transformation.

As with the Portuguese American directorate, the process of creating particular individuals through particular activities cannot be divorced from the impact these activities have on the territory. The two are bound up with each other. In this light, the task force comments that while "Native people" may have a more "holistic approach to forest management, they also recognize the need to maintain high professional standards."[23] High professional standards, it seems, do not mesh with holistic approaches. Just as the "casual method" by which the eighteenth-century "Indian" in Portuguese America obtained the fruits of the forest had to be replaced with a more organized and productive approach, twentieth-century Vancouver Island "Native values" must be adapted to meet "current forest management practices."[24] Whatever skills, occupations, and values "First Nations" people may have had that enabled them to live among the ancient trees of Vancouver Island for centuries without having to clear-cut vast areas on an eighty-year cycle are deemed unacceptable, even "unnatural" to this task force.

The prospect that "First Nations" might receive a favorable ruling on some of their land claims places the British Columbia government in a similar position to that of Portuguese America some two hundred years ago: a (potential) lack of properly disciplined individuals to exploit the resources of the land in a manner that fits with the prevailing economic paradigm. The task quickly becomes one of transforming "First Nations" into "modern nations," that is, making a British Columbian out of a "Native." This practice is still primarily about an ethical orientation to the land.

For that reason, this governmentality cannot be thought in isolation from the practice of creating a territory. Both are ongoing processes that are inextricably linked. In order for the forests of British Columbia to produce particular products marketable in the global economy, a population willing and able to see the forest as a source for such products must be available. Thus planting, spacing, pruning, and fertilizing trees to be harvested every eighty years must become "natural occupations" for members of "First Nations." As these activities become "natural," the forest becomes a tree farm, and vice versa. The circle continues: specific populations are necessary to engage in activities that will transform land into territory, and the act of transforming land into territory will make people into a suitable population. Sovereign territory, including its population, is created through multiple "governmental" practices.

Without a recognition of these practices, explorations of ecopolitical space will continually come up short. The space of ecopolitics is not reducible to geopolitical border lines, or laws, or government buildings. The space of ecopolitics involves practices (very often nonjuridical) that construct both sovereign territories and populations. To repeat: the problem of ecopolitics is not just that the effects of decisions in one sovereign territorial space have impacts on other sovereign territorial spaces; the problem also involves how these sovereign territorial spaces are constructed. In other words, the problem is an ethical one.

ETHICS

The governmental practices that I have focused on throughout this book have taken as a starting point a Lockean relationship between human beings and earth. They have posited that land not used is wasted. They have sought to create both "used" lands and "useful" populations. Sustainable development, in this regard, is no different. It only recognizes limits that Locke did not, or could not, to the material available for development. If we are to come to grips with our global ecocrises, we must come to grips with the ethics of this developmentality, an ethics that is not separate from a discussion of political space.

In a pamphlet from June 1991 entitled "Forest Practices," Mac-Millan Bloedel attempted to align its forestry practices with the World Commission on Environment and Development's notion of sustainable development. Arguing that "most of the coastal old growth is not growing at all [and] in fact, the volume is even declining as a result of rot and other problems," MB justified its "harvesting" of these no-longer-growing forests and replacing them with a farm of trees to be cut down every eighty to one hundred years as an effort to sustain development.[25] Implicit in MB's statement is that these ancient trees, having reached full growth, are going to waste if they are not cut down, since their volume actually starts to decline due to "rot and other problems." The value of the tree is disconnected from the fate of the forest ecosystem. And, moreover, the fate of the forest ecosystem, we are told in this pamphlet, hinges on the

MB staff "developing ideas on how we can retain biodiversity in a way compatible with economic and safe forest management."[26]

The Scientific Panel for Sustainable Forest Practices in Clayoqout Sound, an independent panel convened by the British Columbia government, has articulated a different path for retaining biodiversity. Rather than continuing to place the hopes for biodiversity on economic compatibility, the panel suggests evaluating economic activity in terms of its compatibility with biodiversity. Arguing that "sustaining timber production has historically taken precedence over maintaining forest ecosystems," the panel suggests adopting a "sustainable ecosystem management" approach that would have as its goal not the management of forests as potential products, but as ecosystems.[27]

To a certain extent, I find lines of affinity between this sustainable ecosystems approach and the Nietzschean eco-ethic I articulated in chapter 1. It may prove useful to recall here the contrast I drew between my Nietzschean eco-ethic and Hans Blumenberg's reading of Nietzsche. Blumenberg read Nietzsche as providing a philosophy whereby the world would become material at our disposal. The Nietzsche I read, however, by ceaselessly critiquing the ontological privileging of the place of 'man' in the world and by continually emphasizing the ambiguity, the contingency, and the difference of the world, could not construct a philosophy that would create a world of material at our disposal.

Although the sustainable development position could be made to coincide with Blumenberg's reading of Nietzsche, I cannot see how it could be made to coexist with the Nietzschean eco-ethic I proposed. The sustainable ecosystems approach, however, shares some themes with my Nietzschean eco-ethic, as I have suggested. Both challenge the ontological placement of humanity at the top of a natural order of things. And both place an emphasis on the diversity of nature. Where sustainable development still views natural systems primarily as resources for human consumption and because of that, as we saw with several projects endorsed by the WCED, can operate to reduce biodiversity for the sake of improving production of species deemed useful for development, the sustainable ecosystem management approach seems to challenge the human conceit that all of nature must first of all be for us, whether as raw material or park, and places a priority on the diversity of nature for nature's sake. These are significant differences.

The assertion by the scientific panel that ecosystems be managed will probably raise the ire of many environmentalists. Management, environmentalists have often told us, is human interference; what we need to do is leave nature alone—an echo, in many ways, of Marsh's eco-ethic. The term "management" has been a red-flag term for environmentalists since their emergence. Management seems to imply the opposite of natural. And natural, we all know, means devoid of human artifice. But the extent to which leaving nature alone is now or ever was a feasible option for any species on this planet is something that ought to be questioned. In addition, the binary opposition that is created between management and leaving nature alone is all too close to Locke's opposition of used land and wasteland. A management approach that is centered on sustaining ecosystems could be a fruitful move away from the dichotomy that drives much of both environmentalist and anti-environmentalist discourse.

A focus on sustaining ecosystems that recognizes that all life depends on a diversity of factors and species for survival and sees human beings as one life among many could break the opposition between used land and wasteland, or wilderness and spoiled land. It is here, on the need to break from this ruling opposition, where my Nietzschean eco-ethic comes into direct contact with the issue of governmentality. Although I find this opposition to be wholly artificial, as I hoped to demonstrate through examples of areas once thought to be "natural" that are now recognized to have been "used" by non-European peoples for centuries (chapter 3, primarily), more importantly I find this opposition to be dangerous in that it deflects attention from what I consider the key issue of how differing "ways of life" interact with nature differently. It is this latter issue that lies at the core of my discussion of governmentality.

As I have argued throughout this work, simply recognizing that the space of ecopolitics extends beyond sovereign territory is not an end in itself. Governmentality, to the extent that it is global(izing) today, already thinks community beyond sovereign territory while it continues to construct sovereign territories and the populations that reside within them. The answer is not just seeing that invocations of sovereignty over specific territories are problematic, ecopolitically. Sustainable development, as articulated by the World Commission on Environment and Development, recognizes that. Problematizing the space of sovereign territory without problematizing the ways in which sovereign territories continue to be constructed is only a first step in examining the space of ecopolitics. Ecopolitics is not just about questions of porous geopolitical boundaries; it is about the governmental practices at work that construct those boundaries and the peoples and lands within them. In short, ecopolitics is about governmentality, in particular, a governmentality that sees the earth first and foremost as a resource for human consumption.

What is needed is a new governmentality—one that recognizes that any society must interact with its environment in order to provide for its members and that society must be thought beyond sovereign territorial boundaries, but, perhaps more importantly, one that can think society beyond species boundaries. This need not be taken as an attempt somehow to include nonhumans in human political communities. Just as thinking society beyond sovereign territorial boundaries does not necessitate creating some kind of international governing body to encompass this global human population, but might suggest a smooth or rhizomatic social space, thinking society beyond species boundaries does not necessitate creating some kind of interspecies governing body. Thinking society beyond both sovereign territorial boundaries and species boundaries means taking into serious consideration the ambiguity, contingency, and diversity of life. The problem before us demands no less.

Notes

INTRODUCTION

1. "Brazil Angrily Unveils Plan for Amazon," *Washington Post*, 7 April 1989.

2. Ibid.

3. "Trees, Cows and Cocaine: An Interview with Susanna Hecht," *New Left Review* 173 (January/February 1989), p. 35.

4. World Commission on Environment and Development, *Our Common Future* (Oxford: Oxford University Press, 1987), p. 27.

5. "Global Environmental Power Sought," *Washington Post*, 12 March 1989.

6. Ibid.

7. Ibid.

8. My own use of "our" should not escape interrogation. Most often, when I employ the first-person plural pronoun it will be done as an invitation to those who may consider themselves to be allies of mine. There are places where I employ a rather all-encompassing "we" that is meant to refer to the human species in contrast to a nonhuman nature, or animal existence. There are still other uses of "we" in my text that are meant to highlight how you and I are implicated in many of the ecological crises our planet is faced with today. I am aware that we, as humans, are not all implicated equally, that where we live, how we live, and who we are needs to be considered if responsibility is being handed out, but I have stuck with the term in a belief that many of us feel (erroneously) that how we live our daily lives has no impact on a host of ecological problems around the world today.

9. J. Locke, "Second Treatise of Government," in P. Laslett, ed., *John*

Locke: Two Treatises of Government (New York: Mentor, 1965), p. 339; henceforth referred to as "Locke, 'Second Treatise.'"

10. Ibid., p. 338.

11. Quoted in C. Wagley, "Introduction," in C. Wagley, ed., *Man in the Amazon* (Gainesville: University of Florida Press, 1974), p. 5; emphasis added.

12. M. Foucault, "Governmentality," in G. Burchell, C. Gordon, and P. Miller, eds., *The Foucault Effect: Studies in Governmentality* (London: Harvester Wheatsheaf, 1991), p. 93.

13. For Rousseau's discussion of the difference between sovereignty and government see J.-J. Rousseau, "A Discourse on Political Economy" in C. Sherover, ed., *Of the Social Contract and Discourse on Political Economy* (New York: Harper and Row, 1984). I will elaborate on this discussion in chapter 3.

14. Foucault 1991, p. 93. These variables, depending upon how one reads some of the canonical works in Western political thought, are anything but "mere" up until perhaps the start of the Liberal tradition in the sixteenth or seventeenth century. Both Plato and Aristotle, on up to Machiavelli, it could be argued, took these issues seriously in constructing their theories of politics. Why these issues became "mere" for modern political theorists is perhaps an entirely separate work.

15. A. Diegues, "Social Dynamics of Deforestation in the Brazilian Amazon: An Overview" (Geneva: United Nations Research Institute for Social Development, 1992), p. 10.

16. Quoted in J. Page, "Clear-cutting the Tropical Rain Forest in a Bold Attempt to Salvage it," *Smithsonian* vol. 19 (April 1988), p. 116.

17. D. Quammen, "Brazil's Jungle Blackboard," *Harper's* 276 (March 1988), pp. 65-70.

18. R. Bierregard, et al., "Biological Dynamics of Tropical Rainforest Fragments," *Bioscience,* vol. 42, no. 11 (1992), p. 859.

19. "Study Offers Hope for Rain Forests: Balance of Development and Conservation Proposed," *Washington Post,* 29 June 1989.

1. NATURES, ETHICS, AND ECOLOGIES

1. A. Bramwell, *Ecology in the Twentieth Century: A History* (New Haven: Yale University Press, 1989), p. 2.

2. Ibid., p. 22.

3. Ibid., pp. 22-23, emphasis in original.

4. G. P. Marsh, *Man and Nature,* ed. D. Lowenthal, (Cambridge: Harvard University Press, 1965). For discussions of the historical significance of this text, see C. Glacken, *Traces on the Rhodian Shore* (Berkeley: University of California Press, 1967), J. Passmore, *Man's Responsibility for Nature:*

Ecological Problems and Western Traditions (New York: Charles Scribner's Sons, 1974), Bramwell 1989, and D. Botkin, *Discordant Harmonies: A New Ecology for the 21st Century* (New York: Oxford University Press, 1990).

5. Bramwell 1989, p. 22.

6. Throughout this work, in an attempt to remain consistent with the texts I am reading, I will employ the use of the term *man*—in single quotes—if this is the word used by the author(s) in question.

7. Marsh 1965, editor's introduction, p. xxiii.

8. Ibid., p. 29. For varying discussions of the effects, power, and epistemology of the feminization of nature, see S. Griffin, *Woman and Nature: The Roaring Inside Her* (New York: Harper and Row, 1978); D. Haraway, *Primate Visions: Gender, Race, and Nature in the World of Modern Science* (New York: Routledge, Chapman, and Hall, 1989), and *Simians, Cyborgs, and Women: The Reinvention of Nature* (New York: Routledge, Chapman, and Hall, 1991); E. F. Keller, *Reflections on Gender and Science* (New Haven: Yale University Press, 1985); C. Merchant, *The Death of Nature: Women, Ecology, and the Scientific Revolution* (New York: Harper and Row, 1980); and V. Shiva, *Staying Alive: Women, Ecology, and Development* (London: Zed Books, 1989).

9. For discussions of the conceptions of nature that had been and were being articulated immediately preceding and alongside Marsh, see L. Eiseley, *Darwin's Century* (Garden City, N.Y.: Doubleday, 1970), M. Foucault, *The Order of Things: An Archaeology of the Human Sciences* (New York: Vintage Books, 1973), Glacken 1967, and I. Prigogine and I. Stengers *Order out of Chaos* (New York: Bantam Books, 1984). Marsh's work is definitely "pre-Darwin," in the sense that Marsh still held to a belief that the earth and its inhabitants had not undergone any significant changes in appearance throughout the eons. Marsh believed that the world he explored was much as it had been since creation, save the natural changes that were so slow as to be virtually unnoticeable by 'man,' and the abrupt changes wrought by 'man.'

10. Marsh 1965, p. 36.

11. Ibid.

12. This is *the* question for much of Western environmental thought. Not all environmental authors begin with Marsh's specific premise of a divinely created nature and humanity, but many begin with a belief that at one point in time, humanity's relationship to the earth was much more harmonious. The principal causes given for the destruction of this harmony range from the rise of an anthropocentric religion (see L. White Jr., "The Historical Roots of Our Ecological Crisis" [*Science*, vol. 155, no. 3767 (10 March 1967)]), the "victory" of modern, patriarchal mechanistic science (see Merchant 1980 and Shiva 1989)], and the rise of industrial capitalism

(see M. Bookchin, *Remaking Society: Pathways to a Green Future* [Boston: South End Press, 1990]), to the rise of liberal individualism and its emphasis on acquisition (see W. Ophuls, *Ecology and the Politics of Scarcity* [San Francisco: W. H. Freeman and Co., 1977]), to name just a few explanations.

13. Marsh 1965, p. 36.

14. Ibid., p. 38.

15. Ibid., quoted in editor's introduction, p. xxiv.

16. Ibid., p. 37.

17. Ibid., editor's introduction, p. xxv, emphasis added.

18. Bramwell 1989, p. 23.

19. F. Nietzsche, *The Gay Science*, trans. W. Kaufmann (New York: Vintage Books, 1974), p. 167; henceforth referred to as "Nietzsche *GS.*"

20. G. Deleuze, *Foucault*, trans. S. Hand (Minneapolis: University of Minnesota Press, 1988), pp. 129-30.

21. Nietzsche *GS*, p. 167.

22. Ibid.

23. Ibid.

24. For articulations of mechanistic ecologies, see Ophuls 1977 and P. Ehrlich, *The Machinery of Nature* (New York: Simon and Schuster, 1986).

25. Nietzsche *GS*, pp. 335-56, emphasis in original.

26. F. Nietzsche, *Twilight of the Idols*, trans. W. Kaufmann, in *The Portable Nietzsche* (New York: Penguin Books, 1982), p. 552; emphasis in original; henceforth referred to as "Nietzsche *Twilight.*" Max Hallman provides a translation of this passage in his article "Nietzsche's Environmental Ethics" (*Environmental Ethics*, vol. 13, no. 2 [1991]), which suggests that nature may be the actor here, rather than an assumed human "one": "a high, free, even frightful nature and naturalness, such as plays with great tasks, is *permitted* to play with them" (p. 119, emphasis in original). The shift in action from an assumed human "one" in Kaufmann's translation, to "a high, free, even frightful nature" is significant. This reading of Nietzsche on nature allows for a wider range of agency, and hence reduces the standing of human beings in the world as the only actors. For discussions of the implications of this shift in the realm of agency, see Haraway, 1991 b, in particular "Situated Knowledges," and B. Latour, *The Pasteurization of France, followed by Irreductions: A Politico-Scientific Essay* (Cambridge: Harvard University Press, 1988).

27. Hallman 1991, p. 119.

28. Ibid., p. 100.

29. Ibid., p. 125.

30. Nietzsche *GS*, p. 167.

31. For varying examples of organicism, see M. Bookchin, *Toward an Ecological Society* (Quebec: Black Rose Books, 1980) and Bookchin 1990,

B. Devall, and G. Sessions, *Deep Ecology* (Salt Lake City: Peregrine Smith Books, 1985), and J. E. Lovelock, *Gaia: A New Look at Life on Earth* (New York: Oxford University Press, 1987). There has been a heated debate between Bookchin and his "social ecology" colleagues and the "deep ecologists." While both endorse versions of organicism, they attack one another for the extent to which they include humanity in the organism of nature, or rather, where they place humanity in relation to the rest of nature, with the deep ecologists arguing for a more "biocentric" view. Bookchin presents his critique of deep ecology throughout his 1990 work *Remaking Society*. From my Nietzschean perspective, the social ecology versus deep ecology debate is a theological debate between monotheists (social ecologists) and animists (deep ecologists). The ethic I seek to articulate here breaks from the theology of this debate (hopefully) in its attempt to move beyond the shadow of God.

32. F. Nietzsche, *Thus Spoke Zarathustra*, in *The Portable Nietzsche*, p. 139; henceforth referred to as "Nietzsche *TSZ*."

33. Nietzsche *GS*, p. 168.

34. Nietzsche *Twilight*, p. 517, emphasis in original.

35. J. Granier, "Nietzsche's Conception of Chaos," trans. D. Allison, in *The New Nietzsche*, ed. D. Allison (Cambridge: MIT Press, 1986), p. 139.

36. Nietzsche *GS*, p. 168.

37. Nietzsche *Twilight*, p. 517.

38. Nietzsche *GS*, p. 335, emphasis in original.

39. Nietzsche *Twilight*, p. 484.

40. Nietzsche *GS*, p. 76.

41. F. Nietzsche, *The Will to Power*, trans. W. Kaufmann and R. J. Hollingdale, ed. W. Kaufmann (New York: Vintage Books, 1968), p. 301.

42. Ibid.

43. Nietzsche *GS*, p. 336, emphasis in original. Kaufmann notes that Nietzsche's use of *Dasein*, as with ordinary German, refers to existence in general. He adds that "it is only in Heidegger that *Dasein* refers only to human existence" (p. 336 n. 138).

44. Nietzsche *GS*, p. 278.

45. For an excellent critique of environmentalisms that require "homes" for humanity, see J. Bennett, *Unthinking Faith and Enlightenment* (New York: New York University Press, 1987).

46. For entries into this world of "chaos," see J. Gleick, *Chaos: Making a New Science* (New York: Viking Press, 1987), N. K. Hayles, *Chaos Bound: Orderly Disorder in Contemporary Literature and Science* (Ithaca, N.Y.: Cornell University Press, 1990), Prigogine and Stengers 1984, and C. Froula, "Quantum Physics/Postmodern Metaphysics: The Nature of Jacques Derrida," *Western Humanities Review*, vol. 39, no. 4 (1985). It is no doubt a bit of a stretch to group the various thinkers I will examine next

under one banner, even a banner of "chaos." I will use the term in a broad sense, however, for I find myself at a loss to come up with another way to draw such widely disparate work together without making it all out to be the same. The use of the word "chaos," I think, suggests the variation and difference across the disciplines that I want to maintain; it also maintains a connection to Nietzsche's thought that I would not like to have drift away during this discussion.

47. Prigogine and Stengers 1984, p. 61. Hayles (1990) provides a suitable caution to the use of texts by Prigogine, in particular *Order out of Chaos*. She cautions that Prigogine's work is not considered by many within the scientific community to be "scientific," but is instead seen as far too philosophical or metaphysical, that is, it makes too many conjectures. It is not my intent to cite Prigogine as a scientific authority, a concept that is problematic on its own, but rather, to draw out conceptual challenges that "chaos science" makes to "classical science," both broadly defined. In this respect, *Order out of Chaos* is a powerful source.

48. Prigogine and Stengers 1984, p. 2. Although I don't want to de-emphasize the importance of this shift toward the temporal that Prigogine and Stengers suggest, I do want to point to a danger that may accompany a simple privileging of time over space. If time becomes the element of becoming, of change, of contingency, there is a danger that space will continue to be thought of as static, given, immutable. Edward Soja makes a similar argument in his *Postmodern Geographies: The Reassertion of Space in Critical Social Theory* (London: Verso Books, 1988). Part of my argument in this book is to challenge this static, given conception of space.

49. Gleick 1987, p. 94.

50. The case in question was *Shaw v. Reno* (1993); O'Connor's use of "bizarre" is cited in the notes to L. Guinier, *The Tyranny of the Majority* (New York: Free Press, 1994), p. 266.

51. Ibid.

52. The question of space has been reexamined by a growing number of geographers and social theorists of late. A few have taken up discussions, albeit briefly, of how fractal geometry relates to the new spaces they are investigating. See D. Harvey, *The Condition of Postmodernity* (Cambridge, Mass.: Basil Blackwell, 1989), and Soja 1988.

53. Gleick 1987, p. 65.

54. For an extended discussion of the assumptions and equations of wildlife biology, see Botkin 1990.

55. Prigogine and Stengers 1984, p. 255.

56. Nietzsche *GS*, p. 38.

57. Prigogine and Stengers 1984, pp. 72-73.

58. S. J. Gould, *Wonderful Life: The Burgess Shale and the Nature of History* (New York: W. W. Norton and Co., 1989), p. 11.

59. Ibid., p. 51.

60. Ibid., p. 11.

61. Ibid., p. 44.

62. Ibid., p. 25.

63. Ibid., p. 35.

64. Ibid., p. 44. Gould's use of "accident" calls forth Nietzsche: "Once you know that there are no purposes, you also know that there is no accident; for it is only beside a world of purposes that the word 'accident' has meaning" (*GS*, p. 168).

65. Botkin 1990, p. 9.

66. Ibid., p. 25.

67. Ibid., p. 62.

68. Marsh 1965, p. 29.

69. Botkin 1990, p. 188.

70. Ibid., p. 7.

71. Ibid., p. 10.

72. Nietzsche *GS*, p. 181.

73. Bramwell 1989, p. 10.

74. W. Leiss, *The Domination of Nature* (New York: George Braziller, 1972), pp. 105-13.

75. Such a reading, arguably, can be represented as a "German" one— in keeping with Nietzsche's continual claims that no one read him more poorly than the Germans. This reading, as I have suggested, presents Nietzsche's philosophy as one that champions the rise of modern 'man' to a position from which 'he' will finally be able to take possession of the earth and fulfill the prophecy of an old testament (M. Heidegger, *Nietzsche*, vol. 2: *The Eternal Recurrence of the Same*, trans. D. Krell [New York: Harper and Row, 1984], p. 125). Although Heidegger may be the best-known proponent of such a reading, see also his other *Nietzsche* volumes, as well as *What Is Called Thinking*, trans. J. G. Gray (New York: Harper and Row, 1968), and "Nietzsche's Word: God Is Dead," in *The Question concerning Technology and Other Essays*, trans. W. Lovitt (New York: Harper Torchbooks, 1977). Similar readings of Nietzsche on this issue can be found in J. Habermas, *The Philosophical Discourse of Modernity*, trans. F. Lawrence (Cambridge: MIT Press, 1987), and M. Horkheimer and T. Adorno's *Dialectic of Enlightenment*, trans. J. Cumming (New York: Herder and Herder, 1972).

76. For readings of Nietzsche on the issues of nature, difference, and ethics similar to the one I am offering here, see W. Connolly, *The Augustinian Imperative* (Newbury Park, Calif.: Sage Publications, 1993), G. Deleuze, *Nietzsche and Philosophy*, trans. H. Tomlinson (New York: Columbia Uni-

versity Press, 1983), and S. Kofman, *Nietzsche and Metaphor,* trans. D. Large (Stanford, Calif.: Stanford University Press, 1993). All three could be read as responding to the "German" readings of Nietzsche listed in note 75.

77. Kofman 1993, p. 3.

78. Nietzsche *TSZ,* p. 378; emphasis in original.

79. Deleuze 1983, p. 150; emphasis in original.

80. Nietzsche *TSZ,* p. 379.

81. H. Blumenberg, *The Legitimacy of the Modern Age,* trans. R. W. Wallace (Cambridge: MIT Press, 1985), p. 139.

82. Ibid., p. 137.

83. Ibid., p. 140.

·84. Ibid.

85. Ibid.

86. Ibid., p. 214.

87. Nietzsche *GS,* p. 272.

88. Nietzsche *TSZ,* pp. 162-63.

89. Ibid., p. 161.

2. EXPLORING THE SPACE OF THE INTER(STATE) (I): SOVEREIGNTY

1. See, for example, H. Sprout and M. Sprout, *Toward a Politics of the Planet Earth* (New York: Van Nostrand Reinhold, 1971), R. Falk, *The Endangered Planet: Prospects and Proposals for Human Survival* (New York: Random House, 1971), D. Orr and M. Soroos, eds., *The Global Predicament: Ecological Perspectives on World Order* (Chapel Hill: University of North Carolina Press, 1979), and P. Mische, "Ecological Security and the Need to Reconceptualize Sovereignty," *Alternatives,* vol. 14, no. 4 (1989).

2. R. Keohane, "Neorealism and World Politics" in R. Keohane, ed., *Neorealism and Its Critics* (New York: Columbia University Press, 1986), p. 15. For further discussions of "neorealism," "realism," and their relationship, see other essays in this collection, particularly R. Keohane, "A Theory of World Politics: Structural Realism and Beyond," R. Ashley, "The Poverty of Neorealism," and R. Gilpin, "The Richness of the Tradition of Political Realism," as well as H. Morgenthau, *Politics among Nations* (New York: Alfred A. Knopf, 1948). Morgenthau's works are widely considered the leading statements of realist international political theory.

3. Keohane, "Neorealism and World Politics," p. 13.

4. Ibid.

5. K. Waltz, *Theory of International Politics* (New York: Random House, 1979), p. 80. The speaker in this quotation is anthropologist Meyer Fortes.

6. Ibid.

7. J. Gleick, *Chaos: Making a New Science* (New York: Viking Press, 1987), p. 65.

8. Waltz 1979, p. 8.

9. See F. Nietzsche, *The Gay Science*, books 4 and 5 in particular, for a discussion of these characteristics of life and knowledge. Nietzsche levels a critique against Western philosophy as being all too dependent on finding similarities. For a discussion of the characterization of life depending on difference, see my discussion of chaos theory and ecology in chapter 1.

10. G. Deleuze, *Nietzsche and Philosophy*, trans. H. Tomlinson (New York: Columbia University Press, 1983), p. 101.

11. For examples of such work in global political theory, see R. Ashley, "Untying the Sovereign State," *Millennium*, vol. 17, no. 2, (1988), and "Living on Border Lines: Man, Poststructuralism and War," in J. Der Derian and M. Shapiro, eds., *International/Intertextual Relations* (Lexington, Ky.: Lexington Books, 1989); D. Campbell, *Writing Security* (Minneapolis: University of Minnesota Press, 1992); and R. B. J. Walker, *Inside/Outside: International Relations as Political Theory* (New York: Cambridge University Press, 1992).

12. Waltz 1979, p. 80.

13. Ibid., p. 66.

14. Ibid., p. 68.

15. K. Waltz, "A Reply to My Critics," in Keohane 1986, p. 337.

16. Ibid., p. 341.

17. Ibid.

18. K. Waltz, *Man, the State and War* (New York: Columbia University Press, 1959), p. 2.

19. Waltz 1979, p. 116.

20. Ibid., p. 89.

21. Ibid.

22. See H. Bull, *The Anarchical Society* (New York: Columbia University Press, 1977).

23. Waltz 1979, p. 91.

24. Ibid., p. 88.

25. Ibid., p. 109.

26. For a description of the general debate within international political discourse involving "globalism," see R. Maghroori and B. Ramberg, eds., *Globalism versus Realism: International Relations' Third Debate* (Boulder, Colo.: Westview Press, 1982). R. Falk, *The Promise of World Order: Essays in Normative International Relations* (Philadelphia: Temple University Press, 1987), and M. Soroos, *Beyond Sovereignty: The Challenge of Global Policy* (Columbia: University of South Carolina Press, 1986), offer powerful statements from "globalist" positions, with Falk cautioning

that not all "world order" theorists (himself in particular) advocate a centralized global government, but seek instead to "elevate law and world order values . . . above geopolitical calculations of state advantage" (1987, p. 17).

27. Waltz 1979, p. 210. Waltz speaks of the role "we" must play in solving global problems. The authoritative "we" that governed the reading of Waltz's text becomes the United States at the end.

28. Waltz 1979, p. 66.

29. Ibid.

30. Ibid., p. 113.

31. Ibid., p. 99.

32. William Connolly has explored similarities between these two branches of the international relations tree with respect to the work of Schumpeter and Walzer in his "Democracy and Territoriality," in *Reimagining the Nation,* ed. M. Ringrose and A. Lerner (Philadelphia: Open University Press, 1993).

33. Bull 1977, p. 39.

34. Ibid., p. 8.

35. Ibid.

36. Ibid.

37. Ibid., pp. 8-9.

38. Ibid., p. 9.

39. R. B. J. Walker, *State Sovereignty, Global Civilization, and the Rearticulation of Political Space* (Princeton, N.J.: Center of International Studies, Princeton University, 1988), p. 17. Walker's statement is particularly pertinent to the discussion pertaining to Waltz's theory (or lack thereof) of the state. The silence that surrounds sovereignty in Western political thought allows for thinkers such as Waltz to claim that they are not constructing a theory of the state when they buy into this conception of political space. What Walker has been adept at demonstrating is the extent to which these international political theorists have been theorists of the state. See also Walker 1992.

40. Waltz 1979, p. 66.

41. See D. Garst, "Thucydides and Neorealism," *International Studies Quarterly,* vol. 33, no. 1 (1989), as well as Thucydides, *History of the Peloponnesian War* (New York: Penguin Classics, 1975).

42. See R. B. J. Walker, "The *Prince* and 'The Pauper': Tradition, Modernity, and Practice in the Theory of International Relations," in Der Derian and Shapiro 1989, as well as N. Machiavelli, *The Prince* and *The Discourses,* in *The Portable Machiavelli,* trans. P. Bondella and M. Musa (New York: Penguin Books, 1980).

43. W. Magnusson and R. B. J. Walker, "De-Centring the State," *Studies in Political Economy,* vol. 2 (summer 1988), pp. 54-55.

44. G. Deleuze and F. Guattari, *A Thousand Plateaus: Capitalism and Schizophrenia* (Minneapolis: University of Minnesota Press, 1987), p. 374; emphasis in original.
45. Ibid.
46. Ibid., p. 7.
47. Ibid.
48. Ibid., p. 18.
49. Ibid., p. 12; emphasis in original.
50. Ibid.
51. Ibid., p. 5.
52. Ibid., p. 372.
53. World Commission on Environment and Development, *Our Common Future* (Oxford: Oxford University Press, 1987), p. 27.
54. Deleuze and Guattari 1987, p. 7
55. Quoted in D. Harvey, *The Condition of Postmodernity* (Cambridge, Mass.: Basil Blackwell, 1989), p. 161.
56. Deleuze and Guattari 1987, p. 360.
57. Ibid.
58. Ibid., p. 474; emphasis added.
59. Ibid., p. 384.
60. Ibid.
61. V. Deloria Jr., and C. Lytle, *The Nations Within* (New York: Pantheon Books, 1984), p. 1.
62. A. de Tocqueville, *Democracy in America*, vol. 1 (New York: Vintage Books, 1945), p. 26.
63. Ibid., p. 356. For a further discussion of this general reading of the North American Indians by Europeans, and its effects, see V. Deloria Jr. *Custer Died for Your Sins* (Norman: University of Oklahoma Press, 1976).
64. Deleuze and Guattari 1987, p. 380.
65. De Tocqueville 1945, p. 351
66. Ibid., p. 26.
67. Deleuze and Guattari 1987, p. 386.
68. Ibid.
69. Ibid., p. 357.
70. Deloria and Lytle 1984, p. 9.
71. Ibid., p. 10.
72. Deleuze and Guattari 1987, p. 357; emphasis in original.
73. Ibid., p. 453.
74. Ibid., p. 454.
75. Ibid.
76. Ibid.

77. "The Peru Business Report—advertisement," *New York Times,* 23 November 1993.

78. Ibid.

79. Magnusson and Walker 1988, p. 55.

80. WCED 1987, p. 5. How "seamless" this net is is highly debatable. The WCED's claim of seamlessness operates to level the political playing field across the net. It seems to suggest an equivalence of causes and effects, not just a transversality to them. I will pursue this point further in a more detailed critique of the WCED's *Our Common Future* in the next chapter.

81. Ibid., p. 67.

82. Ibid.

83. Deleuze and Guattari 1987, p. 435.

84. Ibid.

85. See F. Pearce, *Green Warriors: The People and the Politics behind the Environmental Revolution* (London: The Bodley Head, 1991), for a further description of this event.

86. Quoted in *Science* 229 (1985), p. 948.

87. Quoted in Pearce 1991, p. 20.

88. Quoted in ibid., p. 26.

89. Maria Mies discusses the feminist importance of boycotts for creating a connection between women in the "first" and "third" worlds. See M. Mies, *Patriarchy and Accumulation on a World Scale* (London: Zed Books, 1986).

90. B. Barber, "Environment Boycotts Reach Past Governments," *Baltimore Sun,* 13 August 1993.

91. Deleuze and Guattari 1987, p. 454.

3. EXPLORING THE SPACE OF THE INTER(STATE) (II): GOVERNMENTALITY

1. For other "(neo)realist" readings of Rousseau within the field of international politics, see M. Forsyth, H. M. A. Keens-Soper, and P. Savigear, eds., *The Theory of International Relations: Selected Texts from Gentili to Treitschke* (London: Allen and Unwin, 1970), and I. Clark, *Reform and Resistance in the International Order* (New York: Cambridge University Press, 1980). For alternative readings of Rousseau's international political theory, see S. Hoffman, "Rousseau on War and Peace," in his *The State of War* (New York: Praeger, 1965), and M. Williams, "Rousseau, Realism and *Realpolitik,*" in *Millennium,* vol. 18, no. 2 (1989). Both Hoffman and Williams offer convincing readings of Rousseau that contest the (neo)realist reading. My reading of Rousseau differs from theirs in its emphasis on Rousseau's discourse of "government." For Hoffman and Williams, this discourse is apparently unimportant when discussing Rousseau's international political

theory. But, since it adds to the problem of legitimizing sovereign authority within a state, I find it to be crucial to any attempt to cast Rousseau as an international political theorist, be it a realist or otherwise. The reading of Rousseau that I will give here is largely influenced by Michel Foucault's "Governmentality" lecture, printed in G. Burchell, C. Gordon, and P. Miller, eds. *The Foucault Effect: Studies in Governmentality* (London: Harvester Wheatsheaf, 1991).

2. K. Waltz, *Man, the State and War: A Theoretical Analysis* (New York: Columbia University Press, 1959), p. 2.

3. Ibid., p. 231.

4. Ibid., p. 6.

5. J.-J. Rousseau, *A Discourse on Inequality*, trans. M. Cranston (New York: Penguin Books, 1984), p. 111; henceforth referred to as "Rousseau *DOI*."

6. Waltz 1959, p. 168.

7. Rousseau *DOI*, p. 111.

8. Waltz 1959, p. 167.

9. According to Waltz, while the logic of a global Leviathan to control the self-interest of the various states may unassailable, this remedy is "unattainable in practice" (ibid., p. 238). "With each country constrained to take care of itself, no one can take care of the system" (K. Waltz, *Theory of International Politics* [New York: Random House, 1979], p. 109).

10. Waltz 1979, p. 99.

11. Waltz 1959, p. 173.

12. Ibid., p. 174.

13. Ibid.

14. Ibid., p. 175.

15. Ibid., p. 178.

16. Waltz 1979, p. 102.

17. Ibid., p. 103; emphasis added. If Waltz's "we" represents structural theorists of international relations, then perhaps "we" (critics of structural theorizing) can see why "they" easily lose sight of this "fact."

18. Ibid.

19. Ibid.

20. Ibid., p. 104, emphasis in original. The simple opposition of public and private here further suggests the unproblematical re-presentation of the state within Waltz's theory.

21. R. Ashley, "The Poverty of Neorealism," *International Organization*, vol. 38, no. 2 (1984). This essay is edited for R. Keohane, ed., *Neorealism and Its Critics* (New York: Columbia University Press, 1986).

22. Waltz 1986, p. 339.

23. Ibid., p. 340.

24. Recent U.S. military involvement overseas can be read through this lens of domestic violence and legitimation, where the direct force by the U.S. military is not used to respond to "naked aggression" within a sovereign state, but is used when a case can be made for why this aggression extends beyond its state boundaries and affects other sovereign states—particularly the U.S.A. Grenada was presented as a rescue mission largely for U.S. medical students; Panama, a mission of bringing to justice a "known felon" in the war on drugs, at home as well as abroad. The prevalence of such a position, professed respect for the boundaries of the sovereign state, can be seen in recent statements by "just war" theorist Michael Walzer. Walzer draws an analogy between responses to Italy's invasion of Ethiopia in 1937 and Iraq's invasion of Kuwait in 1990: "What we were defending was the idea of a border and the safety and security that everyone enjoys when borders are recognized. The question of democracy comes second. First one defends Haile Selassie on the issue of borders; later one reaches the issue of democracy. But they are separate questions. The attack on Kuwait was a much greater threat to the possibility of *any* world order than is the fact that Iraq or Kuwait is undemocratic . . . There is no security without borders" (M. Walzer, "On Just Wars: An Interview with Michael Walzer," in *Tikkun*, vol. 6, no. 1 [1991] p. 41). For a more elaborate response to Walzer's position see D. Campbell, *Politics without Principle* (Boulder, Colo.: Lynne Rienner, 1993).

25. J.-J. Rousseau, *Of the Social Contract*, trans. C. Sherover, in Sherover, ed., *Of the Social Contract and Discourse on Political Economy* (New York: Harper and Row, 1984), p. 7; emphasis added; henceforth referred to as "Rousseau SC."

26. Waltz 1979, p. 79.

27. Ibid., p. 80.

28. Waltz 1959, p. 175.

29. Rousseau *SC*, p. 15; emphasis in original.

30. Ibid., p. 90.

31. Ibid., p. 47.

32. Foucault 1991, p. 101.

33. Ibid., p. 93.

34. Rousseau *SC*, p. 51.

35. Foucault 1991, p. 93.

36. Ibid.

37. Ibid.

38. J.-J. Rousseau, "Discourse on Political Economy," trans. C. Sherover, in Sherover, ed., *Of the Social Contract and Discourse on Political Economy*, p. 143; henceforth referred to as "Rousseau DPE."

39. Ibid., p. 150.

40. Foucault 1991, p. 101.

41. Ibid.

42. Rousseau *SC,* p. 48. See also the sections preceding this passage, on pp. 36-48.

43. Rousseau *DPE,* p. 154.

44. Ibid., p. 160; emphasis added.

45. Rousseau *SC,* p. 37.

46. Rousseau *DPE,* p. 164.

47. Rousseau *SC,* p. 79.

48. Ibid., p. 80.

49. C. Machlachlan, "The Indian Directorate: Forced Acculturation in Portuguese America (1757-1799)," *The Americas,* vol. 28, no. 4 (1972), p. 359.

50. Quoted in J. Hemming, *Amazon Frontier* (London: Macmillan, 1987), p. 14.

51. List drawn from both Machlachlan 1972 and Hemming 1987.

52. Quoted in Hemming 1987, pp. 11-12.

53. Quoted in Machlachlan 1972, p. 363.

54. Ibid., p. 365.

55. Quoted in Hemming 1987, p. 16.

56. J. Locke, "Second Treatise of Government," in P. Laslett, ed., *John Locke: Two Treatises of Government* (New York: Mentor, 1965), p. 339.

57. Quoted in Machlachlan 1972, p. 364.

58. See M. Rogin, "Liberal Society and the Indian Question," in M. Rogin, ed., *Ronald Reagan, the Movie* (Berkeley: University of California Press, 1987), for a discussion of the nature-Indian-citizen triad in nineteenth-century America.

59. Quoted in Hemming 1987, p. 16.

60. M. Foucault, *History of Sexuality,* vol. 1: *An Introduction,* trans. R. Hurley (New York: Vintage Books, 1980), p. 89; henceforth referred to as "Foucault 1980a."

61. Foucault 1991, p. 95.

62. M. Foucault, "Two Lectures," in C. Gordon, ed., *Power/Knowledge* (New York: Pantheon Books, 1980), p. 104; henceforth referred to as "Foucault 1980b."

63. Ibid., p. 97.

64. Foucault 1980a, pp. 88-89.

65. Ibid., p. 89.

66. Ibid.

67. Foucault 1908b, p. 97.

68. Ibid., p. 98.

69. For discussions of the role of multinational corporations (mnc's)

and multilateral development banks (mdb's) in the development of Brazil, see J. Lutzenberger, "Who Is Destroying the Amazon Rainforest?" *The Ecologist,* vol. 17, no. 4/5 (1987), and S. Hecht and A. Cockburn, *The Fate of the Forest: Developers, Destroyers, and Defenders of the Amazon* (New York: Harper and Row, 1990). Hecht and Cockburn make an excellent argument to the effect that analysis of the destruction of the Amazon rain forest can not be reduced to the activity of mnc's and mdb's, but must take into account the role of the state in seeking to solidify sovereignty over a territory—in short, analysis must take into account the transformative role of government in shaping territory.

70. World Commission on Environment and Development, *Our Common Future* (Oxford: Oxford University Press, 1987), p. 4.

71. Ibid., p. 43.

72. See my discussion of Marsh's *Man and Nature* and the emergence of an ecological perspective in chapter 1.

73. See Hecht and Cockburn 1990 for a discussion of the ecological impact of development projects in the Amazon, and T. de la Court, *Beyond Brundtland: Green Development in the 1990's* (London: Zed Books, 1990), for discussions of the ecological consequences of development around the world.

74. Quoted in J. Bandyopadhyay and V. Shiva, "Chipko: Rekindling India's Forest Culture," *The Ecologist,* vol. 17, no. 1 (1987), p. 33.

75. D. Botkin, *Discordant Harmonies: A New Ecology for the 21st Century* (New York: Oxford University Press, 1990), p. 7.

76. V. Shiva, "Forestry Myths and the World Bank: A Critical Review of 'Tropical Forests: A Call for Action,' " *The Ecologist,* vol. 17, no. 4/5 (1987), p. 142.

77. *Tropical Forestry Action Plan* (Rome: United Nations Food and Agriculture Organization, 1985), p. 41.

78. Quoted in Shiva 1987, p. 142.

79. *TFAP,* p. 85.

80. V. Shiva, *Ecology and the Politics of Survival: Conflicts over Natural Resources in India* (Newbury Park, Calif.: Sage Publications, 1991), p. 336.

81. Ibid.

82. Ibid., p. 169.

83. De la Court 1990, p. 48.

84. Shiva 1991, p. 343.

85. Ibid.

86. Hecht and Cockburn 1990, p. 33.

87. Ibid.

88. Daniel Botkin, in *Discordant Harmonies,* relates a tale about what

was considered to be a "natural" forest in New Jersey. This wooded area had not been cut since at least the early eighteenth century. Hence, it was considered by many ecologists to be an example of a mature forest, a forest that had, over the years, come to achieve a natural balance, a forest that would remain as it was for centuries to come—in short, a natural forest. As it turns out, this hardwood forest was slowly being taken over by softwood trees. This mature forest was changing. What was uncovered about this forest by examining the rings of an old tree felled during a storm was that fires had been periodically set by the "Indians" prior to the European invasion. These fires served to clear out the undergrowth and hence eliminated the softwood trees. The hardwood forest that was the result of this human interference had come to be considered a "natural" forest by many twentieth-century ecologists.

89. G. Esteva, "Regenerating People's Space," in S. Mendlovitz and R. B. J. Walker, eds., *Towards a Just World Peace: Perspectives from Social Movements* (London: Butterworths, 1987).

90. Ibid., p. 292.

91. Ibid., p. 289; emphasis in original.

92. Ibid., p. 292.

93. Ibid., p. 281.

94. Ibid.

95. Ibid., p. 283.

96. Ibid., p. 280.

97. J. MacNeil, *Beyond Interdependence* (New York: Oxford University Press, 1991), p. 20.

98. WCED 1987, p. 5.

99. Ibid., p. 37.

100. Ibid., p. 4.

101. Ibid.

102. Ibid., p. 8; emphasis added.

103. Ibid., pp. 28-29; emphasis added.

104. Esteva 1987, p. 283.

105. A. Escobar, "Discourse and Power in Development: Michel Foucault and the Relevance of His Work to the Third World," *Alternatives,* vol. 10 (winter 1984), p. 394.

106. Esteva 1987, p. 283.

107. Ibid., p. 284.

108. Ibid.

109. Ibid., p. 285.

110. Ibid., p. 284; emphasis in original.

111. Quoted in de la Court 1990, p. 118.

112. Esteva 1987, p. 285.

113. Ibid., p. 288.
114. Ibid., p. 287; emphasis added.
115. Ibid., p. 288.
116. Ibid., p. 287.

4. THEORIES OF ECOPOLITICS: MACHINES, ORGANISMS, CYBORGS

1. A. Dobson, *Green Political Thought* (London: Unwin Hyman, 1990), p. 82.

2. T. O'Riordan, *Environmentalism* (London: Pion, 1976), pp. 303-7.

3. See, for example, B. Ward and R. Dubos, *Only One Earth* (London: Andre Deutsch, 1972), and the World Commission on Environment and Development report, *Our Common Future* (New York: Oxford University Press, 1987). The works listed in this and the following two notes are Dobson's examples unless otherwise indicated.

4. See, for example, W. Ophuls, *Ecology and the Politics of Scarcity* (San Francisco: W. H. Freeman and Co., 1977), and G. Hardin and J. Baden, eds., *Managing the Commons* (San Francisco: W. H. Freeman and Co., 1977). In the Afterword to *Ecology and the Politics of Scarcity Revisited,* Ophuls contests this reading of his work (W. Ophuls and A. S. Boyan Jr., *Ecology and the Politics of Scarcity Revisited* [New York: W. H. Freeman and Co., 1992]). He takes exception to the authoritarian reading that his work has received, and claims instead that it is very democratically oriented (pp. 312-15). I will take up this debate over Ophuls's ecopolitical position in greater detail in the next section.

5. See, for example, R. Heilbroner, *An Enquiry into the Human Prospect* (New York: Harper and Row, 1974), and possibly even E. Goldsmith, *A Blueprint for Survival* (London: Tom Stacey, 1972), although Dobson warns that Goldsmith's suport of participatory democracy may disqualify him from membership in this group.

6. Dobson provides no examples of this subset, but I would suggest works by Murray Bookchin, for example, *Post-Scarcity Anarchism* (Palo Alto, Calif.: Ramparts Press, 1971), *Toward an Ecological Society* (Quebec: Black Rose Books, 1980), and *Remaking Society: Pathways to a Green Future* (Boston: South End Press, 1990).

7. Merchant provides a more comprehensive investigation of the use of these concepts across the field of ecopolitical thought (C. Merchant, *Radical Ecology* [New York: Routledge, 1992]).

8. In particular, I will look at D. Haraway, "A Cyborg Manifesto: Science, Technology, and Socialist Feminism in the Late Twentieth Century" (henceforth referred to as "Haraway 'Cyborg'"), "Situated Knowledges: The Science Question in Feminism and the Privilege of Partial Perspective" (henceforth referred to as "Haraway 'Situated'"), and "Biopolitics and Post-

modern Bodies: Constitutions of Self in Immune Systems Discourse" (henceforth referred to as "Haraway 'Biopolitics'"), all three of which are in *Simians, Cyborgs, and Women: The Reinvention of Nature* (New York: Routledge, Chapman, and Hall, 1991).

9. Ophuls 1977, p. 3.

10. Ibid., p. 144.

11. Ibid., p. 145.

12. W. Ophuls, "Leviathan or Oblivion?" in H. Daly, ed., *Toward a Steady-State Economy* (San Francisco: W. H. Freeman and Co., 1973), p. 222.

13. J. Locke, "Second Treatise of Government," in P. Laslett, ed., *John Locke: Two Treatises of Government* (New York: Mentor, 1965), p. 328.

14. Ibid., p. 333

15. Ibid., p. 332.

16. Ibid., p. 339.

17. Ibid., pp. 338-39; emphasis in original.

18. V. Shiva, *Ecology and the Politics of Survival: Conflicts over Natural Resources in India* (Newbury Park, Calif.: Sage Publications, 1991), p. 168.

19. Ophuls 1973, p. 222.

20. Ophuls and Boyan 1992, p. 313.

21. G. Hardin, "The Tragedy of the Commons," in Hardin and Baden 1977.

22. Ophuls 1977, p. 149.

23. Ophuls 1973, p. 225.

24. Ophuls and Boyan 1992, p. 313.

25. Ibid.

26. Put briefly, the "state-of-nature" analogy follows Hobbes's portrayal of individuals in a state of nature with no authority to keep them in check. Hobbes theorized that these individuals, out of necessity, would have to consent to some form of sovereignty if for no other reason than self-preservation. The analogy has been made that separate sovereign states in the international arena are in a situation similar to that of Hobbes's individuals in the state of nature.

27. Ophuls 1977, p. 210.

28. Ibid., p. 219.

29. Ibid.

30. Ophuls 1977, chapters 5 and 6.

31. Ibid., chapter 7.

32. Ibid., p. 32.

33. Ibid., p. 36.

34. Ibid., p. 32.

35. Ibid., p. 33.
36. Ophuls and Boyan 1992, p. 30.
37. Bookchin 1990, p. 60.
38. Bookchin 1980, p. 67.
39. Bookchin 1990, p. 160.
40. Ibid., p. 66; emphasis in original.
41. Ibid., p. 161.
42. Ibid.; emphasis in original.
43. Ibid., p. 34.
44. Ibid., pp. 45-46.
45. Bookchin 1971, p. 41.
46. Bookchin 1990, p. 187.
47. Bookchin 1971, p. 33.
48. Ibid., p. 34; emphasis in original.
49. Ibid., p. 37.
50. Ibid.
51. Bookchin 1990, p. 120.
52. Ibid., p. 38.
53. Ibid., pp. 35-36; emphasis in original.
54. Ibid., p. 37.
55. Ibid., p. 38; emphasis in original.
56. Ibid., p. 31.
57. Ibid., p. 167; emphasis in original.
58. Ibid., p. 37; emphasis added.
59. Ophuls 1977, p. 32; emphasis added.
60. Haraway "Cyborg," p. 149.
61. Ibid., p. 152.
62. Ibid., p. 150.
63. Ibid., p. 168.
64. Gloria Anzaldúa provides a wonderful description of the border line between the United States and Mexico: "The U.S.-Mexican border *es una herida abierta* where the Third World grates against the first and bleeds. And before a scab forms it hemorrhages again, the lifeblood of two worlds merging to form a third country—a border culture" (*Borderlands/La Frontera: The New Mestiza* ([San Francisco: Aunt Lute Books, 1987], p. 3).
65. Haraway "Biopolitics," p. 218.
66. Haraway "Cyborg," p. 161.
67. D. Harvey, *The Condition of Postmodernity* (Cambridge, Mass.: Basil Blackwell, 1989), p. 164.
68. Haraway "Cyborg," p. 149.
69. Ibid., p. 151.
70. Ibid., p. 154. Here again is an argument for why the chaos theory

examined in chapter 1 is not necessarily salvational. Its role in the creation of this aspect of the cyborg world is far from innocent. For a further discussion of the cyborg aspect of contemporary military technology, see *Cyborg Worlds: The Military Information Society,* ed. L. Levidow and K. Robins (London: Free Association Books, 1989), and M. De Landa, *War in the Age of Intelligent Machines* (New York: Zone Books, 1991).

71. Ibid., p. 151.

72. Ibid., pp. 170-71.

73. Haraway "Situated," p. 195.

74. M. Darnovsky, "Overhauling the Meaning Machines: An Interview with Donna Haraway," *Socialist Review,* vol. 2, (1991), p. 80.

75. Ibid., p. 79.

76. Haraway "Cyborg," p. 150.

77. Gloria Anzaldúa's *Borderlands/La Frontera* (1987) provides an excellent description of this sense of a cyborg existence.

78. Darnovsky 1991, p. 79.

79. Haraway "Situated," p. 191.

80. S. Hecht and A. Cockburn, "Defenders of the Amazon," *Nation,* vol. 248, no. 20 (1989), p. 699.

81. Ibid.

82. Haraway "Cyborg," p. 154.

83. Haraway "Situated," p. 191. The "god-trick," according to Haraway, is the trick of promising objective vision from either the mythic no-where, or the equally mythic every-where. Haraway argues that "objectivity" is only possible from some-where.

84. Ibid., p. 199.

85. Haraway "Cyborg," p. 153.

86. Ibid., p. 152.

5. BRAZIL OF THE NORTH

1. Brazil and Vancouver Island should not be singled out for engaging in this practice. For more on the global devastation wrought by clear-cut logging, see B. Devall, ed., *Clearcut* (San Francisco: Sierra Club Books, 1993).

2. "Progress Report 2: Review of Current Forest Practice Standards in Clayoquot Sound," Scientific Panel for Sustainable Forest Practices in Clayoqout Sound, 10 May 1994, p. 10.

3. R. M. Lee, "Worldwide Campaign against Logging Hits Firm," *Vancouver Sun,* 12 November 1993 (business section).

4. Ibid.

5. Interview with Valerie Langer, 4 July 1994, Tofino, British Columbia.

6. "Who's Buying Clayoquot" (a flyer from The Friends of Clayoquot Sound, 1994).

7. "Profile of British Columbia Exports" (a flyer from Greenpeace Canada, 1994).

8. Ibid.

9. Editorial from the *New York Times,* 6 September 1992.

10. *The Crusaders,* Bonnie View Productions, 16 April 1994. Transcript available from Burrelle's Information Services.

11. *The Crusaders,* 23 April 1994.

12. V. Langer, "Don't Log the Heart out of Clayoquot," *Globe and Mail,* 3 May 1993.

13. Interview with Langer, 4 July 1994.

14. Ibid.

15. Lee, *Vancouver Sun,* 12 November 1993.

16. K. Gram, "Indians Vow All-Out Action to Block Logging Plan," *Vancouver Sun,* 1 May 1993.

17. R. Matas, "Natives Vow to Halt Logging Along Clayoquot Sound," *Globe and Mail,* 1 May 1993.

18. J.-J. Rousseau, "Discourse on Political Economy," trans. C. Sherover, in Sherover, ed., *Of the Social Contract and Discourse on Political Economy* (New York: Harper and Row, 1984), p. 150; emphasis in original.

19. "Native Forestry in British Columbia: A New Approach," Final Report of the Task Force on Native Forestry, submitted to both the Minister of Forests and the Minister of Aboriginal Affairs of the British Columbia Government, November 1991, p. 33.

20. Ibid., p. 34.

21. Ibid., p. 40.

22. Gram, *Vancouver Sun,* 1 May 1993.

23. Task Force on Native Forestry 1991, p. 38.

24. Ibid.

25. MacMillan Bloedel, "Forestry Practices," June 1991, p. 2. Although some might object to my uncritical linkage of MB's statement with the WCED's, I would point to the WCED's endorsement of the Tropical Forestry Action Plan (see chapter 3).

26. Ibid., p. 3.

27. Scientific Panel for Sustainable Forest Practice Standards in Clayoquot Sound, "Progress Report 2," p. 14.

Bibliography

Anzaldúa, G. 1987. *Borderlands/La Frontera: The New Mestiza*. San Francisco: Aunt Lute Books.

Ashley, R. 1986. "The Poverty of Neorealism." In *Neorealism and Its Critics,* ed. R. Keohane. New York: Columbia University Press.

———. 1987. "The Geopolitics of Geopolitical Space." *Alternatives,* vol. 12., no. 4.

———. 1988. "Untying the Sovereign State." *Millennium* vol. 17, no. 2.

———. 1989. "Living on Border Lines: Man, Poststructuralism and War." In *International/Intertextual Relations,* ed. J. Der Derian and M. Shapiro. Lexington, Ky.: Lexington Books.

Bandyopadhyay, J., and V. Shiva. 1987. "Chipko: Rekindling India's Forest Culture." *The Ecologist,* vol. 17, no. 1.

Barber, B. 1993. "Environment Boycotts Reach Past Governments." *Baltimore Sun* (13 August).

Bennett, J. 1987. *Unthinking Faith and Enlightenment*. New York: New York University Press.

Bierregard, R. et al. 1992. "Biological Dynamics of Tropical Rainforest Fragments." *Bioscience,* vol. 42, no. 11.

Blumenberg, H. 1985. *The Legitimacy of the Modern Age*. Trans. R. W. Wallace. Cambridge: MIT Press.

Bookchin, M. 1971. *Post-Scarcity Anarchism*. Palo Alto, Calif.: Ramparts Press.

———. 1980. *Toward an Ecological Society*. Quebec: Black Rose Books.

———. 1990. *Remaking Society: Pathways to a Green Future*. Boston: South End Press.

———. 1991. "Will Ecology Become the 'Dismal Science'?" *The Progressive* (December).

Botkin, D. 1990. *Discordant Harmonies: A New Ecology for the 21st Century*. New York: Oxford University Press.

Bramwell, A. 1989. *Ecology in the Twentieth Century: A History*. New Haven: Yale University Press.

Brown, M., and J. May. 1991. *The Greenpeace Story*. New York: Darling Kindersley.

Bull, H. 1967. "Society and Anarchy in International Relations." In *Diplomatic Investigations*, ed. H. Butterfield and H. Wight. London: Allen and Unwin.

———. 1977. *The Anarchical Society*. New York: Columbia University Press.

———. 1981. "Hobbes and the International Anarchy." *Social Research*, vol. 48, no. 4.

Campbell, D. 1992. *Writing Security*. Minneapolis: University of Minnesota Press.

———. 1993. *Politics without Principle*. Boulder, Colo.: Lynne Rienner.

Chaloupka, B. 1993. "Cynical Nature: Politics and Culture after the Demise of the Natural." *Alternatives*, vol. 18, no. 2.

Chase, A. 1987. *Playing God in Yellowstone*. New York: Harcourt Brace Jovanovich.

Clark, I. 1980. *Reform and Resistance in the International Order*. New York: Cambridge University Press.

Connolly, W. 1988. *Political Theory and Modernity*. New York: Basil Blackwell.

———. 1989. "Freedom and Contingency." In *Life-World and Politics*, ed. S. White. South Bend, Ind.: University of Notre Dame Press.

————. 1991. *Identity\Difference: Democratic Negotiations of Political Paradox*. Ithaca, N.Y.: Cornell University Press.

————. 1993a. *The Augustinian Imperative*. Newbury Park, Calif.: Sage Publications.

————. 1993b. "Democracy and Territoriality." In *Reimagining the Nation*, ed. M. Ringrose and A. Lerner. Philadelphia: Open University Press.

de la Court, T. 1990. *Beyond Brundtland: Green Development in the 1990's*. London: Zed Books.

Crosby, C. 1989. "Allies and Enemies." In *Coming to Terms: Feminism, Theory, Politics*, ed. E. Weed. New York: Routledge.

Crusaders, The. 1994. Bonnie View Productions. 16 and 23 April.

Darnovsky, M. 1991. "Overhauling the Meaning Machines: An Interview with Donna Haraway." *Socialist Review*, vol. 2.

De Landa, M. 1991. *War in the Age of Intelligent Machines*. New York: Zone Books.

Deleuze, G. 1983. *Nietzsche and Philosophy*. Trans. H. Tomlinson. New York: Columbia University Press.

————. 1988. *Foucault*. Trans. S. Hand. Minneapolis: University of Minnesota Press.

Deleuze, G., and F. Guattari. 1987. *A Thousand Plateaus: Capitalism and Schizophrenia*. Minneapolis: University of Minnesota Press.

Deloria, V., Jr. 1976. *Custer Died for Your Sins*. Norman: University of Oklahoma Press.

Deloria, V., Jr., and C. Lytle. 1984. *The Nations Within*. New York: Pantheon Books.

Der Derian, J., and M. Shapiro, eds. 1989. *International/Intertextual Relations: Postmodern Readings of Global Politics*. Lexington, Ky.: Lexington Books.

Derrida, J. 1982. *The Margins of Philosophy*. Chicago: University of Chicago Press.

de Tocqueville, A. 1945. *Democracy in America*. Vol. 1. New York: Vintage Books.

Detwiler, B. 1990. *Nietzsche and the Politics of Aristocratic Radicalism*. Chicago: University of Chicago Press.

Devall, B., ed. 1993. *Clearcut*. San Francisco: Sierra Club Books.

Devall, B., and G. Sessions. 1985. *Deep Ecology*. Salt Lake City: Peregrine Smith Books.

Diamond, I., and G. Orenstein. 1990. *Reweaving the World: The Emergence of Eco-Feminism*. San Francisco: Sierra Club Books.

Diegues, A. 1992. "Social Dynamics of Deforestation in the Brazilian Amazon: An Overview." Geneva: United Nations Research Institute for Social Development.

Dijksterhuis, E. J. 1986. *The Mechanization of the World Picture*. Princeton, N.J.: Princeton University Press.

Doane, M. A. 1989. "Cyborgs, Origins, and Subjectivity." In *Coming to Terms: Feminism, Theory, Politics,* ed. E. Weed. New York: Routledge.

Dobson, A. 1990. *Green Political Thought*. London: Unwin Hyman.

Ehrlich, P. 1986. *The Machinery of Nature*. New York: Simon and Schuster.

Eiseley, L. 1970. *Darwin's Century*. Garden City, N.Y.: Doubleday.

Escobar, A. 1984. "Discourse and Power in Development: Michel Foucault and the Relevance of His Work to the Third World. *Alternatives,* vol. 10 (winter).

Esteva, G. 1987. "Regenerating People's Space." In *Towards a Just World Peace: Perspectives from Social Movements,* ed. S. Mendlovitz and R. B. J. Walker. London: Butterworths.

Everndon, N. 1985. *The Natural Alien*. Toronto: University of Toronto Press.

Falk, R. 1971. *The Endangered Planet: Prospects and Proposals for Human Survival*. New York: Random House.

———. 1987. *The Promise of World Order: Essays in Normative International Relations*. Philadelphia: Temple University Press.

Forsyth, M. 1979. "Thomas Hobbes and the External Relations of States." *British Journal of International Studies,* vol. 5, no. 3.

Forsyth, M., H. M. A. Keens-Soper, and P. Savigear, eds. 1970. *The Theory of International Relations: Selected Texts from Gentili to Treitschke*. London: Allen and Unwin.

Foucault, M. 1973. *The Order of Things: An Archaeology of the Human Sciences*. New York: Vintage Books.

————. 1980a. *A History of Sexuality.* Vol. 1: *An Introduction.* Trans. R. Hurley. New York: Vintage Books.

————. 1980b. *Power/Knowledge.* Ed. C. Gordon. New York: Pantheon Books.

————. 1988. "The Political Technology of Individuals." In *Technologies of the Self,* ed. L. Martin, H. Gutman, and P. Hutton. Amherst: University of Massachusetts Press.

————. 1991. "Governmentality." In *The Foucault Effect: Studies in Governmentality,* ed. G. Burchnell, C. Gordon, and P. Miller. London: Harvester Wheatsheaf.

Friends of Clayoquot Sound. 1994. "Who's Buying Clayoquot."

Froula, C. 1985. "Quantum Physics/Postmodern Metaphysics: The Nature of Jacques Derrida." *Western Humanities Review,* vol. 39, no. 4.

Garst, D. 1989. "Thucydides and Neorealism." *International Studies Quarterly,* vol. 33, no. 1.

Glacken, C. 1967. *Traces on the Rhodian Shore.* Berkeley: University of California Press.

Gleick, J. 1987. *Chaos: Making a New Science.* New York: Viking Press.

Goldsmith, E. 1972. *A Blueprint for Survival.* London: Tom Stacey.

————. 1987. "Open Letter to Mr. Conable, President of the World Bank." *The Ecologist,* vol. 17, no. 2.

Gould, S. J. 1989. *Wonderful Life: The Burgess Shale and the Nature of History.* New York: W. W. Norton and Co.

Gram, K. 1993. "Indians Vow All-Out Action to Block Logging Plan." *Vancouver Sun* (1 May).

Granier, J. 1986. "Nietzsche's Conception of Chaos." Trans. D. Allison. In *The New Nietzsche,* ed. D. Allison. Cambridge: MIT Press.

Greenpeace Canada. 1994. "Profile of British Columbia Exports."

Griffin, S. 1978. *Woman and Nature: The Roaring Inside Her.* New York: Harper and Row.

Guinier, L. 1994. *The Tyranny of the Majority.* New York: Free Press.

Habermas, J. 1987. *The Philosophical Discourses of Modernity.* Trans. F. Lawrence. Cambridge: MIT Press.

Hallman, M. 1991. "Nietzsche's Environmental Ethics." *Environmental Ethics,* vol. 13, no. 2.

Haraway, D. 1989. *Primate Visions: Gender, Race, and Nature in the World of Modern Science.* New York: Routledge, Chapman, and Hall.

————. 1991a. "Cyborgs at Large: Interview with Donna Haraway"; and "The Actors Are Cyborg, Nature Is Coyote, and the Geography Is Elsewhere: Postscript to 'Cyborgs at Large.'" In *Technoculture,* ed. C. Penley and A. Ross. Minneapolis: University of Minnesota Press.

————. 1991b. "Overhauling the Meaning Machines." Interview by M. Darnovsky. *Socialist Review,* vol. 2.

————. 1991c *Simians, Cyborgs, and Women: The Reinvention of Nature.* New York: Routledge, Chapman, and Hall.

Hardin, G., and J. Baden, eds. 1977. *Managing the Commons.* San Francisco: W. H. Freeman and Co.

Harvey, D. 1989. *The Condition of Postmodernity.* Cambridge, Mass.: Basil Blackwell.

Hayles, N. K. 1990. *Chaos Bound: Orderly Disorder in Contemporary Literature and Science.* Ithaca, N.Y.: Cornell University Press.

Hecht, S. 1989. "Trees, Cows and Cocaine: An Interview with Susanna Hecht." *New Left Review* 173.

Hecht, S., and A. Cockburn. 1989. "Defenders of the Amazon." *Nation,* vol. 248, no. 20.

————. 1990. *The Fate of the Forest: Developers, Destroyers, and Defenders of the Amazon.* New York: Harper and Row.

Heidegger, M. 1968. *What Is Called Thinking.* Trans. J. G. Gray. New York: Harper and Row.

————. 1977. *The Question concerning Technology and Other Essays.* Trans. W. Lovitt. New York: Harper and Row.

————. 1984. *Nietzsche.* Vol.2: *The Eternal Recurrence of the Same.* Trans. D. Krell. New York: Harper and Row.

Heilbroner, R. 1974. *An Enquiry into the Human Prospect.* New York: Harper and Row.

Held, D. 1991. "Democracy and Globalization." *Alternatives,* vol. 16, no. 2.

Hemming, J. 1987. *Amazon Frontier.* London: Macmillan.

Hinsley, F. H. [1966] 1986. *Sovereignty.* Cambridge: Cambridge University Press.

Hobbes, T. 1980. *Leviathan.* New York: Penguin Books.

———. 1983. *De Cive: The English Version.* New York: Oxford University Press.

Hoffman, S. 1965. "Rousseau on War and Peace." In *The State of War.* New York: Praeger.

Horkheimer, M., and T. Adorno. 1972. *The Dialectic of Enlightenment.* Trans. J. Cumming. New York: Herder and Herder.

Horowitz, A. 1987. *Rousseau, Nature and History.* Toronto: University of Toronto Press.

Irigary, L. 1991. *Marine Lover of Friedrich Nietzsche.* New York: Columbia University Press.

Jameson, F. 1991. *Postmodernism; or, The Cultural Logic of Late Capitalism.* Durham, N.C.: Duke University Press.

Keller, E. F. 1985. *Reflections on Gender and Science.* New Haven: Yale University Press.

Keohane, R., ed. 1986. *Neorealism and Its Critics.* New York: Columbia University Press.

Keohane, R., and J. Nye. 1977. *Power and Interdependence.* Boston: Little, Brown.

Kofman, S. 1993. *Nietzsche and Metaphor.* Stanford, Calif.: Stanford University Press.

Langer, V. 1993. "Don't Log the Heart out of Clayoqout." *Globe and Mail* (3 May).

———. 1994. Interview (July 4). Tofino, British Columbia.

Latour, B. 1988. *The Pasteurization of France, followed by Irreductions: A Politico-Scientific Essay.* Cambridge: Harvard University Press.

Lee, R. M. 1993. "Worldwide Campaign against Logging Hits Firm." *Vancouver Sun* (12 November).

Leiss, W. 1972. *The Domination of Nature.* New York: George Braziller.

Leopold, A. 1993. "The Land Ethic." In *Environmental Philosophy:*

From Animal Rights to Radical Ecology, ed. M. Zimmerman et al. Englewood Cliffs, N.J.: Prentice Hall.

Levidow, L., and K. Robins. 1989. *Cyborg Worlds: The Military Information Society.* London: Free Association Books.

Locke, J. 1965. "Second Treatise of Government." In *John Locke: Two Treatises of Government,* ed. P. Laslett. New York: Mentor.

Lovejoy, T., ed. 1985. *Amazonia.* New York: Pergamon Press.

Lovelock, J. E. 1987. *Gaia: A New Look at Life on Earth.* New York: Oxford University Press.

Lutzenberger, J. 1987. "Who Is Destroying the Amazon Rainforest?" *The Ecologist,* vol. 17, no. 4/5.

Machiavelli, N. 1980. *The Prince* and *The Discourses.* In *The Portable Machiavelli,* trans. P. Bondella and M. Musa. New York: Penguin Books.

Machlachlan, C. 1972. "The Indian Directorate: Forced Acculturation in Portuguese America (1757-1799)." *The Americas,* vol. 28, no. 4.

MacMillan Bloedel. 1991. "Forestry Practices" (June).

MacNeil, J. 1991. *Beyond Interdependence.* New York: Oxford University Press.

Magnusson, W., and R. B. J. Walker. 1988. "De-Centring the State." *Studies in Political Economy,* vol. 27 (summer).

Maghroori, R., and B. Ramberg, eds. 1982. *Globalism versus Realism: International Relations' Third Debate.* Boulder, Colo.: Westview Press.

Marsh, G. P. [1864] 1965. *Man and Nature.* Ed. D. Lowenthal. Cambridge: Harvard University Press.

Matas, R. 1993. "Natives Vow to Halt Logging along Clayoquot Sound." *Globe and Mail* (1 May).

Mendlovitz, S., and R. B. J. Walker, eds. 1990. *Contending Sovereignties.* Boulder, Colo.: Lynne Rienner.

Merchant, C. 1980. *The Death of Nature: Women, Ecology, and the Scientific Revolution.* New York: Harper and Row.

———. 1992. *Radical Ecology.* New York: Routledge.

Mies, M. 1986. *Patriarchy and Accumulation on a World Scale.* London: Zed Books.

Mische, P. 1989. "Ecological Security and the Need to Reconceptualize Sovereignty." *Alternatives* 14, no. 4.

Morgenthau, H. 1948. *Politics among Nations.* New York: Alfred A. Knopf.

Nash, J., and M. P. Fernandez-Kelly. 1983. *Women, Men and the International Division of Labor.* Albany: State University of New York Press.

Navari, C. 1982. "Hobbes and the 'Hobbesian Tradition' in International Thought." *Millennium,* vol. 11, no. 3.

New York Times editorial. 6 September 1992.

Nietzsche, F. 1968. *The Will to Power.* Trans. W. Kaufmann and R. J. Hollingdale, ed. W. Kaufmann. New York: Vintage Books.

————. 1969. *On the Genealogy of Morals.* In *On the Genealogy of Morals* and *Ecce Homo.* Trans. W. Kaufmann and R. J. Hollingdale, ed. W. Kaufmann. New York: Vintage Books.

————. [1882] 1974. *The Gay Science.* Trans. W. Kaufmann. New York: Vintage Books.

————. 1982. *Thus Spoke Zarathustra* and *Twilight of the Idols.* In *The Portable Nietzsche,* ed. W. Kaufmann. New York: Penguin Books.

Ophuls, W. 1973. "Leviathan or Oblivion?" In *Toward a Steady-State Economy,* ed. H. Daly. San Francisco: W. H. Freeman and Co.

————. 1977. *Ecology and the Politics of Scarcity.* San Francisco: W. H. Freeman and Co.

Ophuls, W., and A. S. Boyan Jr. 1992. *Ecology and the Politics of Scarcity Revisited.* New York: W. H. Freeman and Co.

O'Riordan, T. 1976. *Environmentalism.* London: Pion.

Orr, D., and M. Soroos, eds. 1979. *The Global Predicament: Ecological Perspectives on World Order.* Chapel Hill: University of North Carolina Press.

Page, J. 1988. "Clear-cutting the Tropical Rain Forest in a Bold Attempt to Salvage It." *Smithsonian,* vol. 19. (April).

Passmore, J. 1974. *Man's Responsibility for Nature: Ecological Problems and Western Traditions.* New York: Charles Scribner's Sons.

Pearce, F. 1991. *Green Warriors: The People and the Politics behind the Environmental Revolution.* London: The Bodley Head.

Peterson, V. S. 1990. "Whose Rights? A Critique of the 'Givens' in Human Rights Discourse." *Alternatives,* vol. 15, no. 3.

———. ed. 1992. *Gendered States.* Boulder, Colo.: Lynne Rienner.

Prigogine, I., and I. Stengers. 1984. *Order out of Chaos.* New York: Bantam Books.

Quammen, D. 1988. "Brazil's Jungle Blackboard." *Harper's* 276 (March).

Rogin, M. 1987. "Liberal Society and the Indian Question." In *Ronald Reagan, the Movie.* ed. M. Rogin. Berkeley: University of California Press.

Rousseau, J.-J. 1970. "The State of War." In *The Theory of International Relations,* ed. M. Forsyth, H. M. A. Keens-Soper, and P. Savigear. London: Allen and Unwin.

———. 1984. *A Discourse on Inequality.* Trans. M. Cranston. New York: Penguin Books.

———. 1984. *Of the Social Contract and Discourse on Political Economy.* Ed. C. Sherover. New York: Harper and Row.

———. 1979. *Emile.* New York: Basic Books.

Ruggie, J. 1986. "Continuity and Transformation in the World Polity: Toward a Neorealist Synthesis." In *Neorealism and Its Critics,* ed. R. Keohane. New York: Columbia University Press.

Scientific Panel for Sustainable Forest Practice Standards in Clayoqout Sound. 1994. "Progress Report 2: Review of Current Forest Practice Standards in Clayoquot Sound" (10 May).

Scott, J. "Cyborgian Socialists?" In *Coming to Terms: Feminism, Theory, Politics,* ed. E. Weed. New York: Routledge.

Shiva, V. 1987. "Forestry Myths and the World Bank: A Critical Review of 'Tropical Forests: A Call for Action.'" *The Ecologist,* vol. 17, no. 4/5.

———. 1989. *Staying Alive: Women, Ecology, and Development.* London: Zed Books.

———. 1991. *Ecology and the Politics of Survival: Conflicts over Natural Resources in India.* Newbury Park, Calif.: Sage Publications.

Soja, E. 1988. *Postmodern Geographies: The Reassertion of Space in Critical Social Theory.* London: Verso Books.

Soroos, M. 1986. *Beyond Sovereignty: The Challenge of Global Policy.* Columbia: University of South Carolina Press.

Sprout, H., and M. Sprout. 1971. *Toward a Politics of the Planet Earth*. New York: Van Nostrand Reinhold.

Task Force on Native Forestry. 1991. "Native Forestry in British Columbia: A New Approach" (November).

Thucydides. 1975. *History of the Peloponnesian War*. New York: Penguin Classics.

Tropical Forestry Action Plan. 1985. Rome: United Nations Food and Agriculture Organization.

Vincent, R. J. 1981. "The Hobbesian Tradition in Twentieth Century International Thought." *Millennium,* vol. 10, no. 2.

Virilio, P. 1986. *Speed and Politics*. New York: Semiotext(e).

―――. 1990. *Popular Defense and Ecological Struggles*. New York: Semiotext(e).

Wagley, C., ed. 1974. *Man in the Amazon*. Gainesville: University of Florida Press.

Walker, R. B. J. 1987. "Realism, Change and International Political Theory." *International Studies Quarterly,* vol. 31, no. 1.

―――. 1988. *State Sovereignty, Global Civilization, and the Rearticulation of Political Space*. Princeton, N.J.: Center of International Studies, Princeton University.

―――. 1989a. "History and Structure in the Theory of International Relations." *Millennium,* vol. 18, no. 2.

―――. 1989b. "The *Prince* and 'The Pauper': Tradition, Modernity, and Practice in the Theory of International Relations." In *International/Intertextual Relations,* ed. J. Der Derian and M. Shapiro. Lexington, Ky.: Lexington Books.

―――. 1990. "Sovereignty, Identity, Community." In *Contending Sovereignties,* ed. S. Mendlovitz and R. B. J. Walker. Boulder, Colo.: Lynne Rienner.

―――. 1991. "On the Spatiotemporal Conditions of Democratic Practice." *Alternatives,* vol. 16, no. 2.

―――. 1992. *Inside/Outside: International Relations as Political Theory*. New York: Cambridge University Press.

Waltz, K. 1959. *Man, the State and War: A Theoretical Analysis*. New York: Columbia University Press.

————. 1979. *Theory of International Politics.* New York: Random House.

————. 1986. "A Reply to My Critics." In *Neorealism and Its Critics,* ed. R. Keohane. New York: Columbia University Press.

Walzer, M. 1991. "On Just Wars: An Interview with Michael Walzer." *Tikkun,* vol. 6, no. 1.

Ward, B., and R. Dubos. 1972. *Only One Earth.* London: André Deutsch.

White, L., Jr. 1967. "The Historical Roots of Our Ecological Crisis." *Science,* vol. 155, no. 3767 (10 March).

Wight, M. 1966. "Why Is There No International Theory?" In *Diplomatic Investigations,* ed. H. Butterfield and M. Wight. London: Allen and Unwin.

Williams, M. 1989. "Rousseau, Realism and *Realpolitik.*" *Millennium,* vol. 18, no. 2.

World Commission on Environment and Development. 1987. *Our Common Future.* Oxford: Oxford University Press.

Index

THOM KUEHLS is currently an assistant professor of political science at Weber State University in Ogden, Utah, where he teaches political theory. He received his Ph.D. from the Johns Hopkins University in Baltimore, Maryland. He is the author of "The Nature of the State: An Ecological (Re)reading of Sovereignty and Territory," in *Reimagining the Nation* (Open University Press).